T0229488

RESEARCH INTO SECONDARY SCHOOL CURRICULA

EUROPEAN MEETINGS ON EDUCATIONAL RESEARCH

Part A:
Reports of Educational Research Symposia, Colloquies, and Workshops
organised under the auspices of the Council of Europe

Part B:
European Conferences of Directors of Educational Research Institutions
organised in cooperation with UNESCO, the UNESCO Institute
for Education in Hamburg, and the Council of Europe

PART A: VOLUME 31

COUNCIL OF EUROPE – STRASBOURG

Research into Secondary School Curricula

REPORT OF THE EDUCATIONAL
RESEARCH WORKSHOP
HELD IN MALTA 6 - 9 OCTOBER 1992

Edited by
Paul Heywood
Kenneth Wain
James Calleja

 Taylor & Francis
Taylor & Francis Group
LONDON AND NEW YORK

Library of Congress Cataloging-in-Publication Data

Applied for

CIP-gegevens Koninklijke Bibliotheek, Den Haag

Research

Research into secondary school curricula : report of the Educational Research Workshop held in Malta,
6-9 October 1992 / ed. by Paul Heywood, Kenneth Wain, James Calleja. – Lisse [etc]: Swets & Zeitlinger
– (European meetings in educational research, ISSN 0924-0578 ; v. 31 Pt A,
– Met lit. opg.
ISBN 90-265-1390-9 geb.
NUGI 724
Trefw.: voortgezet onderwijs ; Europa / onderwijsresearch ; Europa

Published by Taylor & Francis
2 Park Square, Milton Park, Abingdon, Oxon, OX14 4RN
270 Madison Ave, New York NY 10016

Transferred to Digital Printing 2007

Cover design: Rob Molthoff

Copyright © 1994 Taylor & Francis

All rights reserved. No part of this publication may be reproduced, stored in a retrieval system, or
transmitted in any form or by any means, electronic, mechanical, photocopying, recording, or otherwise,
without the prior written permission of the publisher.

ISBN 90 265 1390 9
NUGI 724

Publisher's Note
The publisher has gone to great lengths to ensure the quality of this reprint but points
out that some imperfections in the original may be apparent

CONTENTS

Page

Part II: Country Reports

FOREWORD

Michael Vorbeck
Head of the Section for EUDISED and Educational Research
Council of Europe, Strasbourg, France

The Educational Research Workshop on research into secondary school curricula, which took place in Valletta (Malta) on 6-7 October 1992, was organized by the Education Department of the Ministry of Education and Human Resources of Malta in cooperation with the Faculty of Education of the University of Malta, the Foundation for International Studies and the Council for Cultural Cooperation of the Council of Europe (the CDCC).

The Workshop was seen as a contribution to the CDCC's new major Project on "A Secondary Education for Europe".

These educational research meetings regularly bring together research workers and administrators from a selected number of the 37 countries taking part in the work of the Council for Cultural Cooperation. The purpose is to compare research findings and to encourage cooperation in a field of particular interest.

The aims of the Workshop on research into secondary school curricula were:
- to identify and discuss research into secondary school curricula;
- to identify neglected areas of research in this field;
- to indicate how research results can be integrated in secondary school curricula;
- to support cooperation among participating research workers and their institutions.

The following countries were represented: Albania, Austria, Belgium, Cyprus, Czechoslovakia, Denmark, France, Germany, Hungary, Ireland, Italy, Malta, the Netherlands, Norway, Poland, Romania, Slovenia, Spain, Switzerland, the United Kingdom. (Apologies for absences: Estonia, Latvia, Russia).

There were also observers from the Consortium for Development and Research in Education in Europe (CIDREE), the World Confederation of Organizations of the Teaching Profession (WCOTP) and the International Federation of Secondary Teachers (FIPESO).

The list of participants is given at the end of this book.

Six commissioned papers (covering France, Hungary, Malta, the Netherlands, Norway and the United Kingdom) were presented in plenary sessions and then discussed in three working groups. National and individual reports from various countries, as well as lists of research projects and bibliographies, were tabled as background material. On the final day, the Rapporteur General, Dr Clare Burstall from the United Kingdom, summed up the situation and the conclusions emerging from the Workshop.

The Council of Europe is particularly grateful to the Maltese organizers (Dr Paul Heywood from the Ministry of Education and Human Resources, and Prof Kenneth Wain and Dr James Calleja from the University) for their excellent work in preparing and organizing the Workshop and editing the Workshop papers. The Council of Europe would also like to express its thanks to the Rapporteur General, Dr Clare Burstall, to the lecturers and to the group chairmen and rapporteurs.

PREFACE

The Hon. Ugo Mifsud Bonnici
Minister of Education and Human Resources, Malta

Ministers may feel that as they are ultimately also responsible for administration, but more directly and immediately responsible for policy, the principle of subsidiarity is to be applied more strictly and consistently in matters of administration, to a lesser extent in matters of policy. To my mind they (we) should resist the temptation of trying to concoct policy at first hand. The gestation of policy guidelines should involve professional opinion, research, pilot-project experimentation, the stimulation of the first reactions towards achieving consensus. Ministers, however, cannot shirk their responsibility in the final decision-making phase. It is also acquired wisdom that professional opinion will tend to rely on lay - political or mandarin - decisiveness, that research and experimentation may produce ambiguous or ambivalent results, that the favourable echoes from a general public preoccupied with more urgent needs may be slow in arriving, and that the courage of embracing a well developed policy line is a virtue in a man at the helm of a Ministry, albeit not always appreciated except in retrospect.

The proper balance must be achieved and I will be excused if I make a plea for this balance from the awkward position of a Minister confronting experts. The first reflection that I would wish to share with you is that in a practical art such as politics or more prosaically put direction of public affairs, the realizable, the possible, is a constant reining-in factor, to the wildest of ideologies or idealisms. Though some of the practitioners might have harboured the illusion that they would impose a vision, simply by injecting cement into a cloud, time and again experience has shown that there are limits beyond which the most forceful and even the most ruthless cannot go. On the other hand, simple and worn-out experience without a spark of the willingness to change, can be sterile, and sterilizing. Without an utopia, without a striving after a wished-for goal, politics might be reduced to a cynical manipulation of the common goods for very base ends. Although the history of all educational reforms of this century must perforce have a sobering effect on all sprouting reformers, the educational setup, by hypothesis, is *semper reformandum*, in that education which merely reproduces the past is born dead. The politician who is not moved by the inner urge to change must be motivated by baser, not truly political, interests.

The second reflection concerns the interaction between ideological bias and research of actual experience, in the formulation of educational policy. First of all, ideological bias is not monopolized by politicians. The educational professionals have it, the mandarins have it, even where they do not declare it. Politicians have to declare it, indeed to proclaim it. Experience should correct, inhibit and subdue ideology or even idealism, but moral values are not simply the result of research into experience.

Mores do not form morals but though morals can fashion out mores, there is a common human regulating index, which can never be ignored by those who seek to direct public life. Plato and Aristotle and the turn of mind they represent are two points of oscillation not only throughout the whole history of philosophy, but essentially and existentially, in every moment of active political life.

Another reflection concerns the changing fashions or trends in educational policy. The layman may not feel the same instinctual aspiration to conform or be up to date, with the latest *nouvelle vague*. But here again this taking some distance from the professional milieu of educationalists may not be one of the worst traits of Ministers. As a lawyer I did, through long experience in the law courts, come to respect, revere and believe firmly in trial by jury. A fellow jurist, as judge, sums up the case after the two professionals for the parties put forward their case, but the verdict is given by sensible laymen. It is a very sane principle to have experts give advice and opinion but leave the ultimate responsibility for decision to those who have the political mandate. Government by technocrats is not the most responsive to the basic democratic principle of receiving the right to govern from the governed. On the other hand, the knowledgeable and enlightened elected and selected Minister will (hopefully) have the wisdom not to rely on his own expertise and crown himself a technocrat.

Another area which baffles formulators of educational policy is the ambiguity and inconclusiveness of field research or pilot project experimentation. The results can be made to prove what some may be ideologically wanting to prove any way. The viziers on the side of the Minister - the planning councils and think tanks - would be very wrong not to exercise all their caution and caveats.

The particular area of secondary education cries out for fixing of internationally homogeneous standard nomenclature and definition of terms. We need also a clearer definition of role. I am advised that too early specialization or channelling into vocational streams midway is proving counterproductive. Earlier on, when I first assumed responsibility for the Ministry of Education, the decision was taken to assimilate what were then known as trade schools into the Secondary level system. We are now in the process of following through this general line of policy by emphasizing the diagnostic function of pre-vocational schools. The deliberations concerning the postponing of narrow specialization at the immediate post-Secondary stage are also now taking place.

Comparing notes between experts is more important perhaps at this stage than the exchange of opinions between Ministers. This and similar meetings are important in that a common European culture of the function and role of the whole range covered by secondary education within defined bounds is an infrastructure for closer cooperation. Diversity of cultural tradition is a treasure which should be further appreciated but matching of structures and homogeneity of common cores can be a great help in more advanced development.

The cultural wealth of Europe is being further increased by the association to the Council of the nations of the East with all their abundance of variety. In welcoming amongst us not only representatives of the member nations but also our friends from Lithuania and Albania, we are thrilled at the prospect of this cultural enlargement.

I would wish all the experts attending the meeting a happy stay amongst us and may I be allowed to augur well for the fruitful results of your deliberations and the follow-up by Ministers and other servants of the State.

Part One

REPORT

Dr Clare Burstall
Rapporteur Général

European Educational Research Workshop on 'Research into Secondary School Curricula', held in Valletta, Malta, 6 - 9 October 1992.

INTRODUCTION

This Workshop was set up to provide a research input into the CDCC's new major project on 'A Secondary Education for Europe'. The main aims of the Workshop were:

- to identify and discuss research into secondary school curricula;
- to identify neglected areas of research in this field;
- to indicate how research results can be integrated in secondary school curricula;
- to suggest action to be taken by educational policy-makers, teachers and INSET institutions;
- to support cooperation among participating research workers and their institutions.

The Workshop was convened at a time when many of the participants were grappling with the implementation of the dramatic changes brought about by recent political upheavals in Central and Eastern Europe. There was a shared sense of urgency regarding the need to reform the secondary school curriculum in ways that would enhance the life-chances of young Europeans facing the approaching demands of the 21st Century.

The central theme to emerge from the Workshop was a critical review of the role of research in decision-making and in curriculum development and reform, together with an examination of the network of relationships governing the effective exercise of that role.

The report of the Workshop's deliberations is organised as set out below:
1. Research and policy-makers.
2. Research and teachers.
3. Implementation and dissemination of research findings.
4. The secondary school curriculum.
5. The transition period in Central and Eastern Europe.
6. Conclusions and recommendations.

RESEARCH AND POLICY-MAKERS

The quality of the relationship between research and decision-making varies widely from one country to another. In favourable circumstances, research can establish a sound data-base which policy-makers can use to inform their decisions regarding the need for specific educational reforms or curricular innovations. Positive instances of this kind were cited by several countries. In Latvia, for example, both fundamental and applied educational research were seen as supplying a vital input to ministerial decisions on curriculum content, teacher-training programmes, the development of teaching materials, the steps to be taken to raise educational standards, and so on. Similarly, in Cyprus, research findings were used to underpin a number of major ministerial actions, including the setting up of a university, the restructuring of the educational system, the provision of new teacher-training programmes and new teaching materials, and the implementation of a number of curricular innovations and reforms.

In less favourable circumstances, educational reforms may be introduced without prior research or experimentation, but may then be carefully monitored and evaluated, once they are in place. The recent implementation of the 1988 Education Reform Act in England and Wales and the subsequent monitoring of the impact in schools of the new national curriculum and its associated national assessment programme provided an illustration of this approach.

In the worst set of circumstances, changes may be imposed on the educational system arbitrarily, for political reasons, with no prior involvement of either the research community or the teaching profession. Most of the formerly Socialist countries were able to instance examples of this kind of high-handed approach, as were also France and Malta. In each case, there had been no meeting of minds between the policy-makers, the research community and the teaching profession. Decisions were taken for ideological reasons and implemented without advice or consultation; in some cases, they were equally arbitrarily reversed at a later stage, with a consequent considerable waste of human and financial resources.

The Workshop identified a number of issues affecting the quality of the relationship between researchers and policy-makers, namely:

- Researchers and policy-makers operate on different time-scales. Researchers tend to work on a medium to long-term basis, whereas policy-makers tend to operate within the constraints of a short-term political agenda. This important point was made initially in Professor Jacques Colomb's paper: "En effet la différence d'échelle de temps entre le 'temps du politique' qui fonctionne sur le court ou moyen terme et le 'temps de la recherche' qui fonctionne sur le moyen et plus généralement sur le long terme, crée quelques difficultés pour la définition et la prise en compte des résultats des recherches". It was subsequently to recur as an issue on a number of occasions during the Workshop.

- Policy issues are not always amenable to research solutions; research findings are not always directly applicable to policy concerns.

- Research findings may turn out to be critical of current policy and its implementation: policy-makers may react negatively towards researchers if they are perceived as the bearers of bad news. In extreme cases, there may be attempts to undermine the credibility of the research in question or to delay or even suppress the publication of its results.

- Research results may be expressed in difficult and academic language, addressed primarily to other members of the research community, thereby reducing the possibility that their message will reach beyond that community.

- Policy-makers may attempt to use researchers as 'expert witnesses', to buttress a particular political viewpoint, or may choose to heed only those data which tend to support their own position.

- There may be areas of vital importance to policy-makers which have not engaged the attention of the research community. Even the best-intentioned of policy-makers may seek in vain for sound research evidence on which to base urgent educational decisions.

- Policy-makers may be unduly influenced by secondhand accounts of research results, particularly the often brief and vividly-stated versions of them that are presented in the media. Initially misleading impressions can be difficult to counter at a later date.

- Policy-makers are not always prepared to be sufficiently explicit about policy issues which are important to them and may be unaware of the possibility of a potentially productive research input. The research community, for its part, may not be sufficiently alert to current or emerging political trends and their implications for a future research agenda.

The Workshop therefore recommended that:

- Both researchers and policy-makers should seek ways of modifying their respective time-scales, to meet the other's needs with greater sensitivity.

- Researchers should be prepared to make a continuous effort to educate policy-makers regarding both the limitations and the possibilities of educational research. This has to be a genuinely long-term commitment, given the relatively high turn-over that characterises political life.

- Researchers should be prepared to demonstrate to policy-makers how even apparently negative findings can be used to positive effect. The finding that all is not well within an educational system can be used, for instance, to spur remedial action that will reflect well on the makers of current policy.
- Researchers should be more aware of the pressing need to communicate their findings more clearly to a wider audience. They should take pains to highlight both the practical applications of their findings and their implications for the formulation of policy, always in terms readily accessible to non-researchers.

- Researchers should be resolutely and scrupulously a-political and non-partisan in their work and be prepared to combat vigorously any deliberate distortion of their findings for ideological ends. This does not mean, of course, that researchers may not, quite properly, have a political viewpoint of their own, simply that they must be aware of their own value system and not allow it to intrude into their work as researchers.

- As a matter of urgency, the research community should work towards the better documentation of good teaching practices and examples of successful curricular innovations on both a national and international scale. Even in relatively stable and long-established educational systems, there is often little systematic data-gathering and, in countries recently emerging from a totalitarian regime, the situation is much worse and the need for the establishment of national data-bases more urgent. Policy-makers should be prepared to provide adequate financial support for these vital endeavours.

- Policy-makers should avoid undue reliance on secondhand sources of information and be prepared to consult the original research data, if only in summary form. Researchers should bear the information needs of policy-makers in mind when preparing summary versions of their findings.

- Researchers and policy-makers should find ways of working together to produce a mutually satisfactory research agenda. Researchers should be prepared to argue more cogently in favour of the need for fundamental research in certain areas, but should also be prepared to respond positively to requests for shorter-term enquiries. Policy-makers should accept that some forms of research (such as research into the impact of educational reforms, for example) will inevitably require a long-term commitment.

RESEARCH AND TEACHERS

Similar concerns apply to the quality of the relationship between the research community and the teaching profession.

In particular, the Workshop noted the following:

- Research can be perceived as remote from the pressing practical concerns of the classroom teacher.

- Teachers can be impatient for researchers to produce rapid solutions to classroom problems, because they are insufficiently aware of the constraints and complexities of applied research.
- There is a dearth of research that can be used directly by the classroom teacher, often because research findings are expressed in academic terms that do not translate readily into practical applications: teachers do not, on the whole, read research reports.

- Curricular innovations are sometimes developed without teacher involvement and implemented without consultation: they are consequently seen by teachers as being imposed 'from above'.

- There is insufficient discussion between researchers and teachers, particularly during the planning stage of research projects. The early involvement of teachers is essential, if they are to have a sense of 'ownership' of any subsequent classroom innovation.

The Workshop therefore recommended that:

- Where they already exist, links between research institutions and schools should be strengthened; where they do not exist, such links should be established, in order to create more opportunities (including formally time-tabled ones) for the discussion of shared concerns and to work together to set a mutually acceptable research agenda. Researchers should see themselves as working in partnership with teachers, rather than regard themselves as undertaking research 'about' teachers.

- Researchers should make greater efforts to make the results of fundamental research available to teachers, by writing in an accessible manner and highlighting the practical classroom applications of their findings.

- Researchers should acquaint teachers with the inevitable constraints and complexities of long-term research projects, but should also be prepared to show teachers how to carry out their own school-based research and evaluation projects and how to implement their findings in a practical way.

Researchers should seek ways, other than that of the traditional end-of-project research report, of publicising their findings. Possibilities include newsletters addressed to teachers, videos, presentations at teachers' centres, etc.

- Curriculum development and innovation projects should not be undertaken without the active involvement of teachers, from the planning stage onwards. Implementation of curriculum reform should be effected only if the full cooperation of the teachers concerned has been negotiated, appropriate in-service training provided, and adequate support, in both human and financial terms, secured on a long-term basis.

The general view of the Workshop was that it would probably take a generation for a major educational reform to be fully implemented in the classroom, with the attendant danger that some reforms could be ousted by more radical ones, under political pressure, before the former had been given a fair chance to take root.

It was also noted that there was evidence (from France and from Malta, for example) that it was nowadays often the students themselves and their parents who pressed for a return to more traditional approaches, as a result of the threat of unemployment and the related demand for ever-higher qualifications, leading in turn to pressure for more homework, harder assignments, and the rejection of 'extras' outside the mainstream curriculum.

IMPLEMENTATION AND DISSEMINATION OF RESEARCH FINDINGS

These matters have already been touched upon in the two preceding sections of this report. The view of the Workshop, in summary, is that.

- Researchers should express their findings in clear, jargon-free language when they want to reach a wider readership than that of the research community itself.

- Researchers should not hesitate to cultivate the media, and should be prepared to present their findings on radio and television, as well as in newspapers, popular magazines, etc. This implies a commitment to training young researchers in presentational skills, an essential component of their training which is often overlooked: good researchers are not always good natural communicators and need specific training to develop the necessary skills.

- Researchers should avoid the premature release of insecure findings, as well as the introduction into schools of untried innovations, since both of these seriously undermine the credibility of the research effort.

- Both researchers and policy-makers should recognize that the implementation of educational reforms is a gradual and long-term affair and that the teaching profession has the right to expect considerable and sustained support during the implementation process. This support implies not only the provision of adequate human and financial resources, but also the responsibility to ensure that appropriate in-service training is delivered, together with the provision of appropriate and acceptable teaching materials.

THE SECONDARY SCHOOL CURRICULUM

The Workshop recognized the difficulty of defining precisely what the word 'curriculum' should be held to encompass, since it tended to mean different things in different countries: some languages do not even have a word which corresponds closely to the English word 'curriculum'. It was agreed that it would be helpful if a common lexicon could be established for the purposes of international communication.

There was general agreement with the principle of a curriculum that had both a common 'core' and the possibility of a choice of options: it was recognized that even a fairly substantial common area of content could permit a considerable diversity of interpretation. There was some enthusiasm for the idea that the secondary school curriculum should include common elements for all young Europeans.

Discussions concentrated on the following issues:

- The need for a rational basis for the inclusion of elements within the 'core', or, at the very least, the need for a *post-hoc* investigation of their validity.

- The danger that the adoption of a 'minimal' curriculum could eventually lead to minimum standards becoming the maximum aimed at in practice.

- The requirement that local needs and the characteristics of local students should be taken into account in the provision of optional subjects additional to the 'core'.

- The possibility that all common core components should be expressed in cross-curricular terms, embracing wide-ranging skills such as problem-solving, creative thinking, communicative competencies, information-processing, etc. There was, however, general recognition that the implementation of this principle would be an extremely difficult undertaking, given the almost universal supremacy of the subject-based curriculum.

- The need for the common core to leave sufficient space for the exercise of teacher initiative, independence and creativity and the encouragement of local curriculum development. It was felt that these areas of activity had been badly stifled under previously totalitarian regimes and needed very positive nurturing for their effective revival.

- The need for the common core to specify the levels of attainment for students of given ages to reach in the required areas of knowledge, skill and understanding.

There was, however, no consensus on the need or otherwise for the external evaluation and assessment of student achievement. There was a broad spectrum of views, ranging from the extreme standpoint that all external tests or assessments were potentially harmful to students, to the equally extreme opinion that all subjects taught should be formally tested at frequent intervals by standardised 'objective' tests. The middle ground was held by the view that evaluation and assessment were essential components of the teaching-learning process, but that a judicial balance should be struck between informal teacher assessments and more formal external tests. It was noted that several countries, including France, England and Wales, and several of the former Socialist countries, were in the throes of dismantling or de-centralising their former national inspection systems, which threw into sharp relief the need to establish without delay where future responsibilities for school accountability would lie.

- The problem of curriculum overload caused particular concern. There was seen to be a tension between the desire to include new areas of learning, to extend the choice of options, etc., and the need to contain the curriculum within a manageable work-load for both student and teacher.

- The pressure of final examinations, compounded by the threat of possible future unemployment, was felt, in several countries, to distort and narrow the secondary school curriculum and to lead, in a number of instances, to unwelcome premature specialization.

- The issue of interdisciplinarity was a contentious one, and no agreement was reached as to the desirability or feasibility of pursuing this as a realistic goal.

- The idea of the inclusion of a 'European dimension' was generally applauded, but it was felt that there was, as yet, no clear concept of the curriculum content to be involved, the attainment targets to be reached, etc., and that it would therefore be premature to press forward with specific recommendations regarding its place in the secondary school curriculum. The discussion on this issue was closely linked with that on curriculum overload and it was strongly felt that there was a real danger of too many 'new' subjects destroying the coherence of the curriculum. A point emphasised in Professor Jean Ruddock's presentation was that 'coherence' must be understood from the standpoint of the student to whom a particular version of the curriculum was being delivered, not from the standpoint of the curriculum developer.

The teaching of foreign languages and their appropriate place in the curriculum was also discussed at some length, with the particular need to protect and maintain minority languages recognized. It was also felt that the opportunity to study one's own mother tongue should be regarded as a fundamental human right.

It was agreed that as many students as possible should be given the chance to study a variety of foreign languages, as one way of promoting the 'European dimension' in the curriculum. This had clear implications for teachers' initial and in-service training. The particular problem of the re-deployment of former teachers of Russian in Central and Eastern countries was raised, together with the related problem of the scarcity of teachers of other foreign languages in the countries concerned.

Following the presentation of Professor Liv Mjelde's paper, the question of vocational education and its place in the secondary school curriculum was discussed by the Workshop, with particular reference to the issue of cooperative learning and the need to begin the development of necessary generic skills at a pre-secondary level. However, it was felt that there was not sufficient time to discuss these and related issues in depth, and it was suggested that the whole area of vocational education and training might serve as the theme for a future Council of Europe workshop.

There was also considerable discussion about the teaching of humanitarian values. There was some reservation about the desirability of attempting the direct teaching of values at secondary-school level and the question was raised as to whether indirect inculcation of values (through the ethos of the school itself, for example) might not be a better approach to this area. There were also those who felt that the family and the community were more appropriate settings for the teaching of humanitarian values than was the school.

Finally, the Workshop considered the issue of school autonomy, particularly in relation to local curriculum development. It was generally felt that a given degree of school autonomy was an essential pre-condition of effective curriculum reform. Locally-initiated curriculum development was felt to be an essential means of restoring to teachers the independence, creativity and ability to choose amongst competing options that had been largely absent under totalitarian regimes, where the role of the teacher had been down-graded to that of obedient agent. Mr Gabor Halasz's paper described the recent rapid development of school autonomy in Hungary, following the 1985 Education Act, which formally recognized teachers as 'the most important consultative and decision-making body of the school regarding questions of education'. As a consequence of the Act, local or school-level initiatives had become the mainspring of curriculum reform, with new curriculum development reflecting the practical requirements of the classroom as well as the needs and interests of the local population. A similar trend was reported from a number of formerly Socialist countries.

A salutary corrective to this enthusiastic trend towards local school autonomy was provided by Professor Kenneth Wain's paper, which described a system in which the schools enjoyed 'little autonomy or powers to innovate and have virtually no control over the curriculum.' Curriculum development, implementation and assessment were all conducted centrally under the aegis of the Maltese Department of Education, acting on behalf of the Ministry. This example, however, appeared to run counter to the general trend of events.

THE TRANSITION PERIOD IN CENTRAL AND EASTERN EUROPE

The particularly difficult situation of Central and Eastern European countries during the current period of transition from centrally-planned to market economies was felt to warrant special attention.

It was agreed that there was an urgent need for the formerly Socialist countries to develop their own national expertise in research methodology, curriculum design, the development of new teaching materials, including text-books, tests, INSET materials, etc., rather than rely on adopting or adapting the approaches of the West.

It was also agreed that undue deference towards 'foreign experts' should be avoided and that the need to import foreign expertise (in, for instance, training in managerial skills) should be regarded as no more than a short-term expedient. The aim should be to build up the role of the new national research institutions and to establish a body of national expertise in research and development, within the framework of each country's own traditions and culture.

It was felt to be crucially important for developing national research institutions to begin to build up national data-bases, in order to be able to document and monitor the far-reaching changes taking place in their country's educational system, including the sometimes unintended and unexpected consequences of the implementation of educational reforms.

The need to establish an effective dialogue between researchers and policy-makers was felt to be a matter of particular urgency for this group of countries during the current period of transition when rapid changes were being demanded in almost all areas of the educational system and even the establishment of immediate educational priorities required the exercise of a formidable decisiveness, which could be difficult to reconcile with the longer-term research agenda.

It was strongly agreed that the new national research institutions should seek to establish joint ventures, and other forms of collaboration, not only with one another but also with analogous institutions in the West, to share information, research expertise, training opportunities, books, journals, assessment instruments and other materials. This point was forcefully made in Dr Johan van Bruggen's paper and recurred frequently in group discussions.

The possibility of setting up a CIDREE-type organization for Central and Eastern Europe was mooted and received general support, although it was recognized that the new institutions themselves could not be expected to finance this activity in the immediate future. It was hoped that the international funding agencies would be minded to inject the necessary funds to kick-start the operation.

It was clearly recognized that the new national research institutions were labouring under very difficult conditions, which could include limited financial resources (often entailing the need to switch to competitive tendering - a hitherto unfamiliar process - for at least part of their funding), insufficient access to professional literature, inadequate technical support, out-dated communications equipment, and sometimes the presence of ill-motivated or under-skilled staff.

It was agreed that the international research community had a shared responsibility to offer as much support as possible to these institutions, particularly during the difficult early years of the transitional period. In many cases, links that had been forged between institutions in the West and those in Central and Eastern Europe as a result of the Bled meeting of directors of educational research institutions had already proved extremely productive and looked set for a long-term and mutually beneficial relationship.

CONCLUSIONS AND RECOMMENDATIONS

One of the aims of the Workshop was to support cooperation among participating research workers and their institutions. This aim was undoubtedly achieved: it was clear that many already sturdy links were being strengthened during the Workshop and many new ones created, on both an individual and an institutional basis. There was general appreciation of the fact that there was a striking convergence of problems in the educational arena, in spite of the diversity of the educational systems represented at the meeting. It was therefore felt that there was every reason to encourage international cooperation and also a strong incentive to build up a lasting network of mutual support.

The central theme of the Workshop was an examination of the relationships between research and decision-making and between research and educational practice, with particular reference to the reform of the secondary school curriculum and the need for it to reflect the changing face of Europe.

The Workshop's discussions, as the previous pages testify, ranged very widely, but the specific recommendations put forward by the Workshop were confined to its central theme and may be summarised as follows:

- Policy-makers and researchers should attempt a mutually beneficial rapprochement and try harder to respond more positively to one another's needs. They should each be more explicit about the constraints and demands of their respective positions and should strive to communicate more effectively with one another. Although it was clearly recognized that research evidence would never be the sole basis on which political decisions were taken, it was agreed that both parties should work to bring the political and the research agendas into closer harmony.

- Teachers and researchers should work together in a similar way, to increase mutual understanding and promote the more effective practical implementation of research findings. Links between their respective institutions should be strengthened, on a basis of mutual professional respect. Again, the need to communicate more effectively with one another about shared concerns was highlighted: both should aim to work together in a genuine professional partnership.

- Researchers should take every opportunity to improve their communicative skills and to publicise their findings in accessible ways, including relatively unconventional ones. The hallmark of dissemination should be clarity and relevance to the information needs of the intended audience: researchers should make a lifelong commitment to the achievement of this standard.

- The international research community should shoulder its joint responsibility to support the development of the newly-emerging national research institutions in Central and Eastern Europe.

CLARE BURSTALL
Rapporteur Général NFER
25 November 1992

SECONDARY SCHOOL CURRICULA AND EDUCATIONAL RESEARCH IN FRANCE

Jacques Colomb
Director of the Subject Teaching Method Department
INRP, Paris, France

Secondary education in France is split into two separate institutional levels:
- the "Collège" for pupils aged 11-15 years (classes 6 to 3),
- the "Lycée" for pupils aged 15-18 years (classes 2, 1 and "Terminale").

Collèges teach all children in compulsory schooling in a single structure whereas Lycées are divided into three different structural types: general, technical and vocational.

GENERAL ORGANIZATION

The law ("Loi d'orientation") on National Education of 10 July 1989 set out the guidelines for reforms in the French education system for the years ahead.

In respect of the subject to hand, the main point of this law was the creation, under the National Education Ministry, of a National Curriculum Council. The Council has 21 members chosen by the Education Minister for their special qualifications; they are appointed by decree for a five-year term which can be renewed once. The Council issues opinions and puts proposals to the Education Minister relating to the general development and main aims of education. It ensures that the curricula are relevant to these aims and to the way knowledge is developing.

On the basis of guidelines determined by the Minister, draft curricula - dealing with content and method - are produced by groups of experts working on one or more subjects or a given tier of education. These groups include representatives from a wide range of educational backgrounds and report back to the competent departments of the Education Ministry.

This was tantamount to a major overhaul of the procedures for producing curricula, previously the exclusive remit of the General National Education Inspectorate.

The curriculum-setting process is now as shown in diagram 1. Curriculum proposals are drawn up by groups of experts who are selected by the National Curriculum Council according to their special expertise rather than their institutional representativeness. Once adopted by the Minister, these proposals are worked on by the Ministry's pedagogic departments so that they are ready for use in schools, following

Diagram 1. The curriculum-setting process

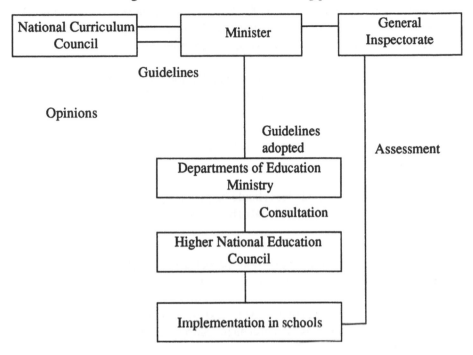

consultation with the Higher National Education Council and a final decision by the Minister. When implemented the curriculum is assessed by the General National Education Inspectorate which reports back to the Minister.

Curricula which emerge from this procedure must be published at least 14 months before their application date to give publishers and authors time to prepare the textbooks for pupils.

This procedure may seem cumbersome but it was designed to ensure greater transparency in the preparation of curricula and greater independence from political interference.

In its work, the National Curriculum Council has taken into account the conclusions of the special topic think-tanks (Commissions de Réflexion Thématiques) set up according to a similar principle in 1988.

These committees have drawn up a number of general and specific conclusions.

General conclusions

There is a need to abandon the traditional approach which ties knowledge up with formal classes, often linked to over-ambitious syllabuses and formal examinations. "Intelligence-based teaching methods" are called for, coupled with an interdisciplinary approach allowing for improved coherence in building knowledge. Whenever possible, experimental and inductive approaches should be preferred to the more usual deductive approach to learning.

Teaching by means of "common cores" and "optional modules" should help to diversify children's school careers as they draw on the children's true motivation.

Teacher training should be genuinely vocational, by integrating both theoretical and practical training in the new training structures through university-level teacher training institutes.

The time to make subject choices should be put back and the three-level course system (general/technical/vocational) should be phased out as it is based on failure.

The concept of assessment, especially at "baccalauréat" level, should be changed, to include an element of continuous assessment.

Moves from one level of schooling to the next (school-collège-lycée-university) should be made smoother, as sudden changes can have harmful effects on children with difficulties.

Conclusions per subject:

Mathematics: "Mathematics for all" is an essential aim and a single scientific "baccalauréat" is apparently vital; "working methods" would be transmitted if the subject were taught on the basis of common cores and modules;

Physics: teaching of physics should be built on experimentation; a determined effort should be made to equip schools for this purpose and to improve harmonious interaction between teaching before and after "baccalauréat";

Biology: the many recent discoveries in this field demand thorough changes to the teaching of this subject and enhancement of both its importance for all children's culture and its essential role in the structuring of scientific knowledge and the acquisition of a critical attitude in experiments;

Chemistry: since chemistry teaching seems to be highly marginalized in France compared with other EEC countries, a significant effort should be made to increase the number of academic contact hours;

Philosophy: from Class 1 onwards, this subject should be integrated into schooling as a way of acquiring knowledge and intellectual methods; it could be taught by grouping subjects together (philosophy/science, philosophy/social science, philosophy/languages/art and literature);

History, geography and social science: these syllabi apparently require some tidying up, cutting out unnecessary elements. Secondary school history syllabi need to be revised (too much 20th century, not enough history of art, religion or science and technology);

French, literature and language: more up-to-date teaching methods, making more use of new technologies, should be introduced; findings have highlighted the need to increase coherence between the teaching of French and foreign languages; civilization and general culture should be given more attention;

Economics: with the dominance of three identifiable aims (cultural, scientific and vocational) in different courses, the current situation is unclear and a complete rethink is called for in order to achieve a better integration of the teaching of economics into education as a whole.

On the basis of these conclusions reached by committees and the work done by the various groups of the National Curriculum Council, the Council has drawn up the main guidelines for secondary education and proposed these to the Minister of Education.

GUIDELINES FOR COLLÈGES

The report by the National Curriculum Council (November 1991) proposed new guidelines for developing the Collèges.

At present, the fact that Collèges teach all children of the same age poses certain problems essentially linked to how to cope with the growing heterogeneity of school children.

The "Loi d'Orientation" makes Collèges responsible for giving all pupils the tools they need to understand and control the personal, social and civil situations they will have to face and to start thinking about the career they would want to aim for.

This aim is an ambitious one, demonstrating a desire to make the necessary cultural and economic progress called for as the year 2000 dawns. But it is also a difficult aim to achieve.

The proposals made for the objectives of Collèges focus on three areas:

Subject skills which express the aims to be achieved in each subject vis-a-vis the general aims of the Collèges;

Methodological skills which must be learned through specific instruction;

Fundamental attitudes which highlight the values a democratic society, based on solidarity, needs if it is to progress properly. Such attitudes must be developed through all subjects.

This places particular emphasis on the need for changing the aims of Collèges, traditionally centred on the goals of each subject. Such a new approach will prove problematic both in terms of the form and substance to be given to these new learning processes which depart from the usual subject-related standards.

The National Curriculum Council proposes ways of bringing about these necessary changes based on two key elements:
- reference to a list of aims to be reached in the Collèges, expressed in terms of both subject and methodological skills, included in the syllabi;
- the fact that classes are now no more than heterogeneous reference groups, according to pupils' needs and their immediate environment.

The question of guidance is also of vital importance for the Collèges. It is often in this age group that guidance begins and has essentially been based on failure. Current proposals are intended to correct this trend, shifting the emphasis to more positive guidance, upgrading technological or tertiary training as compared with traditional education, based primarily on formal intelligence.

These different objectives have been translated into basic guidelines, according to the various subjects:
- teaching French: the top priority at the Collèges is to enable all pupils to "master the French language as a tool of written and oral communication". Teaching of French must therefore:
- confront school-children with various situations of oral or written communication and provide them with the means to cope with them;
- construct and exercise, through these practices, the language skills which form the basis of mastery of the mother tongue.

According to the National Curriculum Council this entails:
- concentration on texts without separating reading from writing activities;
- diversifying the types of writing practised in clearly-defined communication situations;
- linking the learning of grammar to production tasks;
- constructing oral learning around clearly-defined situations;
- not limiting the teaching of French to French classes alone.

Developed in this way, the teaching of French should therefore ease access to culture, for example, and not neglect the teaching of literature.

Teaching of technology: this is the second fundamental guideline for Collèges. As well as the specific aims of the teaching of this subject (comprehension, study, making technological products) there is the added importance of this subject for guiding pupils and the part it can play in building up notions of different jobs.

Relevant proposals include the gradual introduction of technology into Collèges starting with the "technology of discovery as a part of the introduction to science and technology" in classes 6 and 5; then, in classes 4 and 3; more hours would allow pupils to see how technologically talented they are, with career guidance in mind.

Teaching of experimental science: this instruction is often wrongly specialized and compartmentalised by academic reference subjects (physics, biology, chemistry, etc) and should be re-thought as part of a subject realignment, linking science to technology.

The experimental aspects of the sciences, largely ignored, should now be developed. However if these experiments are to work properly, some practical measures need to be taken.

Teaching of modern languages: introducing children to modern languages at primary school is an experimental measure which needs to be assessed as regards its impact on the Collèges. In any case, developing the teaching of modern languages should be a priority in the years ahead.

"Transverse themes": consumerism, development, information, health, security, environment, etc, are often neglected because of the problems inherent in the compartmentalisation of subjects at school. The Council considers it necessary to set aside a special interdisciplinary slot for classes on these subjects.

Alongside this "traditional learning" which places most emphasis on a formal approach to knowledge, the Council suggests the introduction of scientific and technical workshops aimed at developing actual abilities and manual/practical skills, familiarisation with technical objects and the ability to take decisions, etc.

Along similar lines, the Collèges should also introduce training in documentary research and the use of computers as a tool for receiving, producing and communicating information, in text or audiovisual form.

These guidelines have not yet been reflected in a reform of the Collèges, which is still on the agenda to be carried out in the near future.

GUIDELINES FOR LYCÉES

The reform of the Lycées will come into force at the beginning of the next academic year (1992-93). It is based on various proposals put forword by the National Curriculum Council and accepted by successive education ministers (Lionel Jospin and Jack Lang).

It breaks new ground in two main areas: firstly by introducing three hours per week of "personal learning classes"to help pupils with difficulties or provide advanced teaching for the most gifted pupils. This is a way of responding to the problems inherent in dealing with the growing heterogeneity of Lycée classes, linked to the growing diversification of pupils, as the Lycées will soon be receiving 80% of all schoolchildren at "baccalaurèat" level.

This personalized learning assistance, almost completely ignored by the Lycée until now, will be part of normal instruction. At the beginning of the next school year (1993-94) Lycée class 2 will include modular lessons in mathematics, French, history/geography and modern languages. Pupils will be placed in groups according to the findings of a national assessment designed to determine their needs. In the time set aside for this purpose teachers will be able to draw on their own subject knowledge to provide subject or methodological assistance for pupils with difficulties, provide for interdisciplinary or transdisciplinary learning or give extra tuition to brighter pupils. Special syllabuses will only be used in "Terminale", whereas in Classes 2 and 1 curriculum choices will be left to the teacher.

The second main area affected by reform of the Lycées is the development of a simpler, more coherent system of courses, allowing pupils to express their choices rather than be forced to follow outside guidance as is the case at present. The new system will offer three general courses: literature, economics/social studies and science; and three main technology courses: science and industrial skills, science and tertiary skills, medico-social skills. These new more clearly profiled courses ought to fulfil their aims and not stray from their original roles as is the case at present.

Unless it is derailed by the system, this new approach to guidance, centred on the pupil's personal career plans marks a major turning-point in the conception of studies. Democratization of the Lycées demands a change, with Lycées adapting to all types of pupils and promoting a cultural community uniting pupils, teachers and the knowledge which still has to be developed.

The need for coherence in the teaching of different subjects, the need for a transdisciplinary approach to certain kinds of learning, the recognition of the importance of methodological approaches to learning, the much greater importance attached to the pupil's own plans and the need to consider new assessment arrangements are the main features of the new guidelines for Lycées. These guidelines clearly dovetail with the proposals for the reform of Collèges.

EDUCATIONAL RESEARCH AND ITS PLACE

When curricula are being prepared, research findings are obviously taken into account, but only in a very incidental, non-institutionalized way. The different time scales between "politician-time" - which is short or medium-term - and "research-time" - which is medium or more often long-term - make it more difficult to define and take account of research findings.

In spite of these difficulties, the impact of research can be identified, as can neglected areas to which attention needs to be paid. In respect of curricular problems research takes two main directions:
- research into subject-teaching methods, looking at questions relating to the definition and the transmission of knowledge in the education system;
- research into how education works, more directly linked to psychology, sociology or, more usually, education science.

In this report, only the former, research into teaching methods, will be dealt with as it lies at the heart of the problem to tackle.

Research into subject-teaching methods has been taking place in France for some 15 years and it is only now that it reaches maturity in certain subjects (mathematics, French, experimental science, etc). It still has a lot of ground to cover in other subjects. The concepts and methodologies gradually being worked out make it possible to approach fundamental problems linked to knowledge which lie at the heart of the definition of the curricula.

This report will briefly examine some major questions on which research into teaching methods has produced a certain amount of significant knowledge on the basis of which the problem of defining and applying the curricula might be tackled. I have opted firmly for a systems approach in which education phenomena will be analysed in terms of knowledge flows through what Chevallard calls the "didactic system" (a system composed of the "pupil-teacher-knowledge" trio and bilateral relations between each of its three components).

Seen in this way knowledge shifts position dramatically according to the level examined.

SCHOOL KNOWLEDGE occupies the central place in the system (See diagram 2). It is knowledge socially identified at a particular moment in time as requiring to be taught in a given context.

One of the merits of research into teaching methods has been to scrutinize this school knowledge, to bring to light its origins and to put an end to considering it as sacrosanct.

Diagram 2

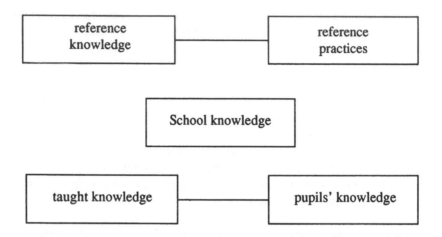

These origins vary widely but are always characterized (regardless of whether we mean knowledge or practices) by the transformation they must undergo in order to make them possible to teach.

The phenomena of "didactic transposition", brought to the fore by Chevallard, in mathematics teaching methods in order to analyse the shift from "scholarly knowledge" to "knowledge to be taught", have opened the way for a whole range of research projects on teaching methods adding up to a breakthrough in terms of the anthropology of knowledge.

This approach, almost exclusive to mathematics, could be extended to other subjects by considering other origins of school knowledge:
- reference practices introduced by Martinand in the field of experimental science
- that knowledge which is a pure creation of the school system.

By applying the concept of didactic transposition to other subjects, much research work has revealed the complex relationships between the various types of knowledge and practices that go to make up school knowledge, in a way that normally defies easy comprehension.

Epistemological analysis of these different types of knowledge has brought to light in various subjects some of the ways in which school epistemologies work but about which hardly anyone ever asked questions.

Knowledge of teaching methods produced in this way is a powerful tool for analysing many educational phenomena observed inside and outside the classroom, although the processes of transposition occur completely upstream of teaching itself.

This epistemological analysis is just one aspect of the research work into teaching methods being developed in close connection with another side, concerned with the transmission of this same school knowledge in the education system.

TAUGHT KNOWLEDGE - or school knowledge as "stage-managed" by the teacher in his own way - is the most visible part of the way the didactic system works.

The visibility does not make this knowledge anymore directly understood.

Various works on teaching methods have led to the invention of tools which enable some of the processes involved in this "textification" of knowledge to be grasped.

Teachers, here, are at the centre of the teaching phenomena by dint of the way they control two fundamental elements which ensure that teaching works properly:

management of teaching time and its interaction with learning time;

responsibility for the meaning of knowledge and for how much of it is devolved to the pupils themselves.

The amount of progress in building up knowledge which pupils make depends heavily on both of these variables.

The quality of pupils' knowledge varies according to the choice made by the teacher when organizing learning situations and making sure they lead from one to another.

A teaching approach in which the teacher is dominant places pupils in a position of receiving knowledge already built by the teacher whereas a "constructivist" approach places them in a position of building knowledge by themselves.

In both these extreme cases, responsibility for the meaning of knowledge passes from the teacher to the pupil, each in its own way, but neither model seems more efficient than the other, contrary to the very widely held view which currently tends to favour the latter approach.

In the same field, the theory of teaching situations devised by Brousseau has highlighted the fundamental role of didactic variables which are manipulated, consciously or not, by the teacher when "stage-managing" knowledge.

PUPILS' KNOWLEDGE has also been the object of numerous studies in teaching methods which have shed some light on how this vital part of the system works.

Work on pupil representations has led to the formulation of teaching approaches in which pupils cease to be seen as "empty boxes" in which knowledge is to be poured but are treated instead as subjects already possessing knowledge which, albeit often erroneous or incomplete, needs to be developed.

Numerous studies of pupils' mistakes have succeeded, through a thorough analysis of their causes, in pin-pointing some of the obstacles which crop up in teaching time and time again.

Consequently, one promising approach involves making mistakes an element on which learning can be built.

Removing the stigma of being "wrong" from making mistakes profoundly alters pupils' relationship with knowledge. The consideration being given to new AS-SESSMENT arrangements could be a powerful and by no means negligible way of altering teaching practices.

Curricula in most subjects have taken on board many of the findings of this research. Taking the long term into account in learning, granting a new status to mistakes, paying attention to personalizing learning processes and the consideration given to methodological and transverse learning approaches are just some examples of the effects this research has had on curricula.

Given the ever-growing pace of developments at present, this research must be continued as so much ground remains to be covered.

This area of educational research needs to be extended much further and systematized within a more clearly defined curricular framework. This should be in line with more clearly defined pedagogical structures and methods of assessment.Research should focus on producing the necessary tools needed for managing teaching and training. This would greatly affect curriculum development and teacher training.

Research on curricula might well be a priority in the years ahead, although it is not the case at present.

CURRICULUM CHANGES IN SECONDARY EDUCATION IN HUNGARY

Gábor Halász
Head of Research Department
National Institute of Public Education, Budapest

THE DEFINITION OF SECONDARY EDUCATION

It is perhaps not useless to start the presentation of curricular changes in secondary education with a definition of what exactly secondary education is, since the meaning of this term seems to vary from one country to another. In Hungary the term 'secondary education' normally designates education offered by schools receiving pupils beyond the age of fourteen, after the completion of the eight-year basic school. Nevertheless, by the term "secondary school" ("kozépiskola") most people understand only the four-year general schools ("gimnázium") and the four or five-year professional schools ("szakközépiskola") offering preparation for the maturity examination ("érettségi") and do not place in this category the shorter three-year vocational schools ("szakmunkásképzo") which prepare students for the skilled-worker certificate.

In Hungary the four higher grades of the eight-year basic school ("általános iskola"), are not considered as secondary level education. Recently the borderline between basic and secondary education has been negligible as some secondary schools started receiving pupils before the age of fourteen. This was made possible by a 1990 amendment to the Education Act in force which allowed secondary schools to set up lower classes. Consequently a growing number of pupils enter secondary education without going through the last grades of the eight-year basic school (at 10 or 12). Another new trend is the setting up of ninth or tenth classes by some eight-year basic schools in order to absorb those pupils who have not been admitted to a secondary institution.

The present horizontal and vertical structure of the system is shown in annex 1.

THE PATTERNS AND LEGAL FRAMEWORK OF CURRICULUM CHANGES

Up to the eighties the almost exclusive source of curriculum change in Hungarian schools was overall periodical reforms affecting the entire system. The last curriculum reform of this type, leading to the introduction of new curricula at nearly all educational levels and school types, was carried out in 1978-79.

A new feature of the preparations of the 1978 curriculum reform was the active participation in the committee work of leading scholars from the Academy of Sciences. Nonetheless, the whole process was still controlled by the party political apparatus and ministerial bureaucracy urging rapid execution. Although the presence in the process of progressive academic circles resulted in the creation of a number of modern, progressive programmes, the bureaucratic organization and the hasty implementation of the reform led to serious anomalies. In fact, despite the significant positive changes in several subjects (especially in mathematics, linguistics, literature and the sciences) and a general modernization of the content of teaching brought about by the reform , as a whole it was immediately perceived by many experts as a complete failure.

The main causes of the 1978 curriculum reform failures have been identified as follows:

- the new curricula elaborated by experts working in central institutions were introduced in several subjects without serious experiment and there were no satisfactory consultations with practising teachers. As a consequence, the requirements set by them were not sufficiently adapted to the pupils' real capabilities and, in certain subjects, even the minimum requirements were calculated so that they could not be achieved by the average pupil. As certain educationalists later remarked, the pedagogical transformation of the academically good programmes was not achieved.

- the reform was implemented without sufficient financial resources as it started in a period of growing cuts in public expenditure. Consequently the necessary materials and textbooks were not produced in time and in an adequate quantity. Teachers were not given appropriate supplementary training so they were not prepared to use the new materials.

- soon after the implementation of the new curricula the working days of the week were reduced from six to five which made the planned-time schedules unusable.

- although originally one of the main goals was that of offering alternative curricula and increasing differentiation, this was not brought about. Therefore the former rigidity in the curriculum and time schedules was perpetuated.

However, the reform brought about two important changes opening certain possibilities for differentiation. First, the hitherto unified requirements were divided into two categories: minimum ones and supplementary ones. This gave teachers greater freedom to adapt their teaching to the abilities and needs of their pupils. Second, in the last two grades of secondary schools (16-18 age group) optional subjects were introduced in both the theoretical and the practical fields.

According to the 1978 reform the time schedule of general secondary schools (gimnáziums) was fixed as shown in annex 2.

The perceived failure of the 1978 reform gave an impetus to the development of ideas advocating gradual and locally initiated changes instead of overall central reforms, and differentiation instead of unification. A new development paradigm began taking shape which led to the radical revision of existing curriculum theories and administrative arrangements.[1]

The idea of having differentiation and school level development as the main focus of curriculum regulation was confirmed by the 1985 Education Act.[2] One of the main objectives of the Act was the enlargement of school level autonomy. As it was formulated by the Act: "in the framework of statutory provisions and central educational programmes" schools might define "their own educational tasks", elaborate "their own local educational system" and "work out supplementary curricula". Individual schools were given the jurisdiction of deciding all issues related to their organization and work that were not relegated by statutory provisions to the authority of other agencies. The Act defined the teaching staff as "the most important consultative and decision-making body of the school regarding questions of education". Schools were asked to prepare their own pedagogical programme and "local educational system" which was to be approved by the teaching staff after consultation with the representatives of external agencies.

A fundamental change resulted from the fact that local and regional authorities formally lost their right to interfere in curriculum matters. Although the new Act was not followed by substantial changes in the existing central curricula, the possibility for local change was considerably broadened by the radical reduction of external control on school operations. The system of school inspection run before by 19 country authorities was suppressed and new advisory centres (county pedagogical institutes) were set up in order to offer professional counselling services to schools. Advisors were encouraged to use modern evaluation methods instead of the old paternalistic model based on the control of individual teachers. These changes, in fact, abolished the organizational basis of central pedagogical direction and separated pedagogical orientation from legal administration.

As a result of these changes, a few years following the new Act, the number of school level innovations started increasing in a dramatic manner. In 1988 a central innovation fund (with about 200 million Hungarian Forints, i.e. approx. 16,000 US $ to be distributed every year) was created, allowing individual schools to apply for financial support in the event of initiating school level curricular or organizational changes. By the end of 1991 approximately one quarter of the schools applied for some support from this fund.

CURRICULUM CHANGES IN THE GENERAL SECTOR

Since the beginning of the eighties two kinds of curriculum changes can be identified: in the first part of the decade centrally initiated counter-reforms were made in the curricula of several subjects; in the last years of the decade local or school level initiatives became the main source of change.

Since the shortcomings of the 1978 reform were soon perceived, as early as the beginning of the eighties the correction of the most apparent anomalies was started. The counter-reforms were elaborated in a changing institutional framework: under the control of professional circles with decreasing involvement of political and administrative structures and with the increasing involvement of practising teachers. In general the counter-reforms resulted in the following changes:

- as the result of longer and deeper coordination with practising teachers curricula became more realistic and better adapted to the needs of school practice;

- the proportion of optional elements and possibilities for choice has increased;

- the ideological character of subjects such as history or philosophy was reduced; a general depolitization of curricula had started.

As mentioned, while before the 1985 Education Act most reforms were introduced at system level, after the Act, most of them were initiated from the school level. Almost all experts agree that the present Hungarian system is characterized by an increasing diversity. At present there is no reliable record on locally initiated curriculum changes but it is certain that their number is very high.

The most important curricular changes in the general (non-vocational) sector after the 1978 reform were the following:

- Owing to the reduction in the number of working days, modifications were made in the timetables. Some requirements were shifted from the minimum category to the supplementary one.

- By the middle of the decade, in certain subjects, alternative curricula were drawn up; in other fields, such as computer studies, entirely new curricula were produced. In the last years the number of alternative curricula and textbooks has increased remarkably.

- In the general sector specialized tracks, suppressed by the 1978 reform, were gradually reintroduced. In 1990 there were 165 specialized classes using particular centrally elaborated curricula (about half of the Gimnáziums resort to this option). Another form of specialization was offered by the facultative curricula of the higher classes. In 1990 there were about 15 officially approved optional curricula leading to formally acknowledged qualifications.

- School level differentiation was dramatically speeded up by a special provision of the 1985 Act which allowed schools to apply "particular solutions". The most frequent "particular solutions" have been the introduction (1) of "human" and "real" tracks, (2) of special foreign language classes, (3) of classes in computer studies (4) of integrated subjects. In Spring 1992 Ministry officials estimated that up to the beginning of the year, about one fourth of secondary schools had introduced some approved curriculum changes.

- After 1987 the most important change was the setting up of bilingual Gimnáziums. In these schools pupils receive intensive language training in a one-year preparatory class ("O" form) and, in the upper grades, certain subjects are taught in the foreign language. This development received special financial support from the government even in periods of budget cuts. A number of professional secondary schools also started bilingual classes.

- Fundamental curriculum changes might be generated by the appearance of six-year and eight-year Gimnáziums. Curricula and manuals used by these schools have been worked out generally by the teaching staff themselves, and have been directly transferred (often sold) by one school to another. So far no central curriculum has been elaborated for these types of schools. According to expert opinions on the approved curricula of these new school forms, the following changes can be observed: (1) in the six-year schools a 4 + 2 division is frequently used with increasing specialization in the second part. In the eight-year schools the 2 + 4 + 2 division is typical with a tendency of introducing integrated, child-centred curricula in the first part and growing specialization in the third one; (2) Many of the eight-year Gimnáziums return to the traditional pre-war curricular models: they introduce Latin and classical subjects, and lay special emphasis on language teaching. Interestingly, in the state sector most schools returning to the eight-year pattern are rather low quality institutions using the structural changes to improve their quality.

- The most radical curriculum changes have appeared in the new private schools. In fact, curriculum changes introduced by private institutions can be characterized by three different tendencies: (1) academic orientation with an effort to offer efficient preparation for higher education; (2) pupil-centred orientation with progressive pedagogical methods; (3) practical orientation with special attention to the handicapped and those threatened by unemployment.

- Important curricular changes can be expected in schools transferred back to the churches, with the revival of traditional teaching styles, the introduction of classical subjects and religious education. So far the churches have not shown signs of activity in the field of curriculum development. In general, the quality of teaching is seen as high in most church secondary schools, a few of them being especially innovative.

CURRICULAR CHANGES IN THE VOCATIONAL SECTOR

As to the vocational sector, the different tracks have different curriculum problems. In fact there are three types of vocational schools leading to the "maturity" examination (see annex 3). In schools leading to technician qualification, no professional theoretical subjects are taught in the first grades, while in the other types these subjects take about one fifth of the time schedule in the same grades. In the third and fourth grades of schools, preparing for skilled worker qualification, pupils have to do professional practice (i.e. work) on two days a week, while they do only one day a week in the two other types.

In fact, the negative consequences of the rigidity of central curriculum control were first perceived in the vocational sector. Economic organizations introducing new technologies often complained about the impossibility of introducing new training forms in schools as this depended on long and slow processes of central approval. The rapid changes in the economic context made it necessary that the adaptation of the content of training to the economic needs be speeded up. By the second half of the eighties a new principle emerged in training policy: that of adaptability. This meant the acceptance of the idea that the introduction of new training forms must be more and more dependent on direct agreements between schools and economic organizations.

By the end of the eighties, school-level curriculum development became the typical form of curriculum change also in the field of vocational training. This was particularly hastened by a World Bank development project which raised a special fund for those schools which introduced school-level curriculum development. Schools were given the possibility of applying for special subventions with concrete plans for new training courses. In fact, at present, it is rather central coordination and professional support for these local developments which is seen as lacking. New training courses have been opened mainly in the fields of trade, banking and informatics. Many vocational schools started intensive language teaching courses, some of them trying out the teaching of certain subjects in a foreign language.

The economic crisis had dramatic influence on the three-year vocational training schools leading to the skilled worker certificate. Since practical training for pupils enrolled in these schools had often been assured by contracted firms, the bankruptcy of the latter entailed the disappearance of the training place itself. Many three-year vocational schools were obliged to devise or adapt new courses of general professional preparation.

In 1991 new central curricula were elaborated in order to allow schools to open special transition classes. The new curricula contain three larger blocks: general science course (12 hours a week in most programmes), social skills (2 hours) and professional courses (18 hours). At the end of 1991, there were 15 eligible professional courses, all of them consisting of two parts: a three-month professional orientation course and a seven-month specialized course. Most programmes for these special courses were modelled on German programmes of the same type. Most special classes have been opened as the ninth or tenth grades of the eight-year basic school but many of them have started in the three-year vocational schools.

RECENT DEBATES AND PROSPECTS

Recently curriculum policy has become one of the central issues of educational debate in Hungary. At the heart of these debates is the question of curriculum control. The radical advocates of output control reject all kinds of central curriculum defining compulsory content and are in favour of controlling teaching exclusively through external assessment and examinations. Others think that the control mechanisms prior to the 1985 Education Act should be restored with detailed definition of the content of teaching for all school types and grades.

An answer to this question can be found in the draft of a National Core Curriculum (NCC) elaborated by expert committees and published in Spring 1992. The work was initiated by a group of curriculum and evaluation experts in 1989 and, although it received a certain support from successive governments, it has never gained official status. Despite the lack of official status, and the reluctance of central administration to accept it as such, at present there are no alternative proposals for curriculum policy. After its publication the NCC was sent by the Ministry of Education to all Hungarian schools as a document submitted to public debate. Thus, while it is under discussion and deviates fundamentally from the official curricula, the NCC is now playing a leading role in orienting teaching activity together with the older official documents yet in force.

In fact, the NCC is an attempt to adapt curriculum regulation to the current conditions of the Hungarian educational system characterized by increasing school level autonomy.[3] Its main characteristics are the following:-

(a) It is based on the principle of two-level regulation combining a centrally defined frame-curriculum and achievement requirements with locally chosen or elaborated detailed curricula. The NCC is defined as a central frame-curriculum which - contrary to the earlier central curriculum - is not meant to regulate directly classroom level teaching activity. This latter function is supposed to be monitored by the different alternative curricula, programmes and textbooks published by both public and private agencies.

(b) It is based on the principle that the control of curriculum development must be independent of the vertical structure of the system, that is the central frame-curriculum must allow various local structural arrangements. The NCC is supposed to be followed by schools organized into different structural patterns as 8+2+2, 6+4+2,

4+4+2 etc. The NCC defines content and requirements only for the 6-16 age group, that is for the compulsory period. The teaching in higher classes is supposed to be regulated by the examination system ('maturity' and vocational examinations).

(c) It is based on the principle that instead of concrete subjects the central frame-curriculum must define broader knowledge areas. Actually, the NCC contains eleven knowledge areas: mother tongue, foreign languages, mathematics, visual and mass communication, informatics, science, man and society, arts, technics and environment, family and housekeeping, physical education.

The reluctance of the central administration to accept the NCC as an official document is due to the radical innovation of these principles. The acceptance of these principles would certainly make it rather difficult for the central administration to directly control the teaching processes in schools, but in fact, in the present administrative context (with almost all schools owned by autonomous self-governments and with enlarged school level freedom) the possibilities for direct central control are in many ways restricted. Another source of uncertainty is probably linked to the undoubtedly modern and liberal spirit of the NCC which is in contradiction with those political forces in favour of promoting traditional and conservative values in schools.

For the time being one of the greatest problems for the development of teaching is the lack of a sufficient number of alternative textbooks and programmes. Although during the last year a high number of alternative programmes have been produced by innovative teachers and different authors, the lack of adequate financial support and distribution mechanisms impedes the emergence of a working textbook and programme market. This problem will probably be partly solved by the expected introduction of a new system of finance which would transfer state financial support for textbooks directly to the schools instead of the state publishing house. The number of private or smaller public publishing houses entering the textbook market has remarkably increased during the last two years but at present they are not able to offer reasonable prices.

Another often discussed problem is that of evaluation mechanisms. During the last few years the idea of establishing a centrally coordinated examination system and introducing standardized assessments for certain age groups was developed by different educational circles. In certain universities and institutions for professional support new evaluation units have been set up. These have developed modern evaluation methods creating the institutional foundations for regular assessments. However, the lack of appropriate financial and political support hinders an adequate functioning of these institutions.

THE EFFICIENCY OF THE SYSTEM IN THE LIGHT OF SURVEYS

The lack of appropriate evaluation mechanisms is seen as one of the weaknesses of the Hungarian educational system. Since the beginning of the seventies, when Hungary joined the International Association for Educational Assessment (IEA), several surveys have been conducted allowing the global evaluation of the efficiency of teaching in Hungarian schools. These surveys, together with the expertise in curriculum theory and planning imported through this cooperation, had a considerable impact on curriculum development.[4]

According to the IEA surveys the quality of science teaching is especially high in Hungarian schools. In 1970 from 15 countries, Hungary held the second rank position for the age group 14 (last grade of the eight-year basic school) and the seventh one for the age group 18 (last grade of the four-year secondary school). In 1983 from 25 countries it held the first position for the age group 14; while from 19 countries it held the third position in Physics, the fourth in biology and the seventh in Chemistry for the age group 18. Reading achievement, however, seems to be constantly low.[5]

Recent monitor surveys also show a light decrease of achievement of 14-year-old pupils. From 1986 to 1991 reading test results have decreased by approximately 8%, while results in mathematics are 5-7% lower.[6]

Notes

1. Báthory, Z., 1986 "Decentralization Issues in the Introduction of the New Curriculum: the Case of Hungary," *Prospects.* Vol. XVI, No. 1.
2. Halász, G., 1987 "A New Education Act,"*The New Hungarian Quarterly,* Summer , Vol. XXVIII, No. 106.; and: (from the same author) "The reform of educational administration in Hungary," *Prospects,* Vol. XX. No. 3, 1990.
3. Nagy, J. - 1990 Szebenyi, P., *Curriculum Policy in Hungary,* Hungarian Institute for Educational Research, 1990.
4. Halász, G. & Lukács, 1986 P., *The Impact of the IEA in Hungary,* manuscript, Hungarian Institute for Educational Research.
5. Báthory, Z., Tanulók, 1992 "isolák, különbségek," (Pupils, schools, differences), *Tankönyvkiadó*, Budapest, 252-263. pp.
6. Az Országos Közoktatási Intézet értékelési központjának jelentése a tanulók tudásszintjéröl (The report of the evaluation center of the National Institute of Public Education on the achievement of pupils). *Pedagógiai Szemle*, Vol. XLII., No. 4., 1992., 3-20. pp. PSZLE, cikk.

ANNEX 1

The horizontal and vertical structure of the Hungarian educational system

Normal age

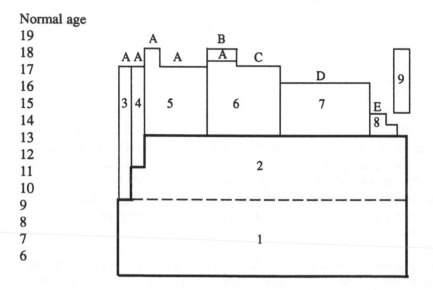

Index

1. Lower classes of the eight year general school (class teaching)
2. Upper classes of the eight year general school (subject teaching)
3. Eight year extended Gimnázium
4. Six year extended Gimnázium
5. Four year regular Gimnázium bilingual schools with "0" classes included
6. Professional secondary schools (PSS) with three tracks (all offering maturity exam. plus prof. qual.)
    - leading to technician qualification (4+1 year)
    - leading to secondary professional qualification
    - leading to secondary professional qualification and skilled worker qualification
7. Vocational schools (VS)
8. Shorter vocational education and special classes (SC)
9. Adult education

A   maturity examination
B   technician qualification (one year following the maturity exam)
C   maturity exam. plus secondary professional qualification
D   skilled worker qualification
E   lower level vocational qualification

ANNEX 2

Weekly time schedule of general secondary schools according to the 1978 reform (hours)

Subject	I.	II.	III.	IV.
Hungarian language	2	2	1	1
Hungarian literature	2	3	3	3
History	2	2	3	4
Philosophy	-	-	-	2
Foreign language	7	5	3	2
Mathematics	5	4	3	3
Physics	2	2	3	3
Chemistry	2	4	-	-
Biology	-	-	4	2
Geography	3	2	-	-
Music	1	1	1	-
Arts	1	1	1	-
Physical Education	3	3	3	3
Technique	2	2	-	-
Form-teacher's lesson	1	1	1	1
Career orientation	-	1	-	-
Compulsory optional	-	-	7	9
TOTAL	**33**	**33**	**33**	**33**

Typical weekly time schedule of vocational secondary schools leading to maturity examination (hours)

Subject	I.	II.	III.	IV.	V.
Hungarian language and literature	3	3	3	3	-
Foreign language in schools leading to ordinary secondary qualification	2	2	2	2	-
Foreign language in schools leading to technician qualification	4	4	optional		-
History	2	2	2	2	-
Mathematics	4	4	3	3	-
Science	5	5	depends on field		-
Philosophy	-	-	-	2	-
Physical education	2	2	2	2	2
All general subjects	**18-20**	**18-20**	**12-14**	**12-14**	**2-4**
Vocational theoretical subjects					
In schools leading to skilled worker certificate or to ordinary secondary qualification	5-8	5-8	6-10	6-10	-
In schools leading to technician qualification	-	-	5-9	5-9	20-22
Vocational practice					
In schools leading to ordinary secondary qualification	5-7	5-7	5-7	5-7	-
In schools leading to technician qualification	6-8	6-8	6-8	6-8	6-8
In schools preparing for skilled worker qualification	5-7	5-7	12-14	12-14	-
Optional subjects	-	-	0-2	0-2	-
All subjects together	**30-32**	**30-32**	**30-32**	**30-32**	**30-32**

RESEARCH AND THE SECONDARY SCHOOL CURRICULUM

Jean Ruddock
Director, QQSE, University of Sheffield

The introduction of a *national curriculum* as part of the 1988 Education Reform Act has been the most significant influence on the secondary school curriculum. The national curriculum, and related legislation, will be the major focus of this paper. First, however, it is important to outline the structure of educational research in England and Wales.

THE STRUCTURE AND STATUS OF EDUCATIONAL RESEARCH

Let a hundred flowers blossom and let a hundred schools of thought contend. (Mao, 1962)

There is no official *research* institution supported directly and fully by the government. Government Departments have their own teams of in-house statisticians, researchers and advisers, but the Departments, in addition to carrying out some in-house research, increasingly invite selected individuals/teams to tender for research contracts, or they set up an open competition for designated research moneys. Some resources for educational development and evaluation have, in the context of the recent educational reform, also been made available by some new government agencies - the National Curriculum Council (NCC) and the School Examinations and Assessment Council (SEAC).

Research is also supported by other categories of institution, some of which receive funds from the government but which nevertheless have a measure of independence.

First, there are the major Research Councils, of which the Economic and Social Research Council (ESRC) funds research in education. The ESRC can give money for research centres or programmes (which operate over a number of years); it can fund initiatives which are made up of a number of linked research projects; it can fund series of research-focused seminars and it can fund individual projects. In each case, proposals are submitted and judgments are made through a careful process of peer review. The Council also has a system for monitoring and evaluating the research that it supports.

Second, there are 'specialist' organizations such as the Equal Opportunities Commission and the Commission for Racial Equality which have limited budgets but which nevertheless support some small scale research.

Third, there are a number of independent organizations which fund educational research, some of which have reasonably large research budgets at their disposal - such as the Leverhulme Trust, the Rowntree Trust, the Gulbenkian Foundation. These tend to have clear priorities (the Gulbenkian Foundation, for instance, is particularly interested in the arts and the Leverhulme Trust has an interest in industry, education and the economy).

There are also research and development initiatives supported by industry. For example, British Petroleum (BP) has been prominent in its support for education, particularly - but not exclusively - in relation to the secondary science curriculum.

The picture, then, is of a diversity of sources, agencies and emphases in educational research - hence the opening quotation by Chairman Mao. One possible effect of this diversity is lack of clear influence. Indeed, the introduction of the national curriculum is an example of the way in which a government which has been in office for some time is able to move towards legislation on highly controversial policy issues and on the basis of a "consultation" procedure which has low validity and low credibility and without taking much account of research.

The research community recognizes the problem, as the following quotations - from researchers of quite different educational perspectives - acknowledge (all the writers are discussing the introduction of the national curriculum):

> ... it was palpably clear that the educational research community was ignored by politicians and by policy makers ... The voice of the educational researcher is not heard ... except in those instances where it is convenient to those who determine events to quote "research" to justify policies which they are determined to pursue in any event ... (McNamara, 1990)

> ..., many educational researchers feel that the principal proposals of the Education Reform Act have been presented to the public and to Parliament with insufficient account being taken of educational research. (Homan, 1991)

And perhaps the strongest statement:

> It is not just that educational research played little or no role in shaping the Act, but rather that in many respects the legislation goes in opposite directions to those indicated by the findings of educational research. (Hammersley, 1992)

But, Hammersley goes on, this is not just a recent state of affairs: 'Indeed, if one looks back at the history of British educational research since the mid-century, I think one has to conclude that research has rarely had much influence (on national policy)'. He argues for a fundamental rethinking of the relationship between research and educational practice (in which, I assume, he includes the practices of policy makers).

McNamara proposes that educational research needs to be, and to be seen to be, more consistently interested in 'the essential purpose of schooling' - ie teaching and learning. There may be something in this analysis but one must also bear in mind Hammersley's point that 'the knowledge produced by research' is not any longer 'a God's eye view' but rather 'a perspective from a particular angle, one whose appropriateness can be challenged' (Hammersley, 1992). Such relativism makes it easier, of course, for government, if it wishes, to justify not taking research seriously.

Grace's solution (1991) is based on his analysis that educational researchers have 'for too long ... spoken to each other ... in a technical and arcane language'. He goes on: 'The socio-political and economic conditions of the present time require us to speak more effectively and more accessibly to the wide community ... using a good sense which is based upon careful, objective and critical scholarship and research in education'. His answer lies in the quality of the research that we undertake - but he also assumes that research should focus on topics that are of major educational significance and recognized to be so by both politicians and the public.

Thus, the organization, resourcing and status of educational research in relation to the practice of teachers and policy makers is a central issue for those of us concerned with schooling and the curriculum.

RECENT CHANGES TO THE SECONDARY SCHOOL CURRICULUM

While the main focus of this paper is the national curriculum and the related programme of assessment, it is important that the reforms be seen in the context of the Government's wider policy for education. The main features are these: consumer choice and an increased diversity of kinds of secondary schools; the reduction of democratic local control of schools in favour of more centralized control and greater autonomy for individual schools; more accountability, for schools, for the quality of the education they offer; more information to be made public about the achievements of individual pupils and individual schools; a new system of regular inspection of schools and new arrangements for dealing with schools judged to be failing their pupils.

It is also important to see the reforms against the background of curriculum development in the last two or three decades. Secondary schools in England and Wales have been relatively free to determine their own curriculum. The last period of major reform came in the 60s and 70s when the post-Sputnik crisis of knowledge led, through the curriculum development movement, to the creation of alternative subject-based 'projects' which gave schools considerable choice of content and approach. Exhilarating as many of the products of this movement were, there was no overall guidance that could ensure curriculum coherence across schools or across the different curriculum areas within any one school.

The new national curriculum should give both coherence and continuity of learning. Its main aim however 'is to raise standards of achievement and the quality of teaching' (NCC document). It consists of three core subjects: English, maths and science; seven foundation subjects: technology (including design), history, geography, music, art, physical education and, for pupils between 11 and 16, at least one modern foreign language. These ten subjects constitute the basis of a curriculum which will cover 'the range of knowledge, skills and understanding commonly accepted as necessary for a broad and balanced curriculum'. The curriculum must also include opportunities to study particular "themes" and "dimensions" - some in a cross-curricular way: health education and education for citizenship; career education and guidance; economic and industrial understanding; environmental education; concern for gender and multicultural issues; political and international understanding. The ten basic subjects are expected to take up about 80 per cent of curricular time and if space exists beyond the study of themes and dimensions, then pupils can choose to take, say, an additional language.

Programmes of study have been designated for each of the core and foundation subjects and attainment targets are being drawn up, each with ten attainment levels. A structure of "Key Stages (KS)" provides an organizational structure for learning and a rough guide to expected attainment:

KS 1 - first 2 years of schooling ages 5-7
KS 2 - next 4 years of junior schooling ages 7-11
KS 3 - first 3 years of secondary schooling ages 11-14
KS 4 - last 2 years of compulsory schooling ages 14-16

The intention is that pupils' attainments will be judged by externally prescribed standardised assessment tasks or tests at the end of each Key Stage, with the public examination at 16 being brought into line, in time, with the attainment targets. The results of assessments are to be reported for each Key Stage and there has been, through the drawing up of a "Parents' Charter", considerable emphasis on informing parents about the achievements of individual pupils and of the school as a whole.

Schools are relatively free to plan the proportion of time given to the defined elements and to choose how to organize coverage of the themes and dimensions. They are also free to decide on teaching and learning styles, materials and text books. There are no centrally required or recommended texts.

In short, the national curriculum and its related assessment programme 'represents the attempt by government to redefine educational standards ... in terms of subject-specific criteria and attainment targets which are ... organized hierarchically in terms of levels, the teaching of which is to be ensured by regular testing' (Torrance,

1991). It is recognized that if the Government is committed to raising educational standards (and accepts that this means more than raising test scores) then the curriculum *and* the tests must reflect 'good educational practice and afford students the opportunity to develop the capacity to think, solve problems, work collaboratively and so forth' (ibid).

The first reported assessments of maths and science for Key Stage 3 (ie those that affect the secondary school curriculum) will take place in summer 1993 and for English and technology in summer 1994.

A two-stage programme of monitoring and evaluation for the core subjects and for technology was launched in 1989. Stage 1 focuses on the monitoring of the first year of implementation of each subject in order to "identify matters of concern". Stage 2 is "a more sustained and detailed investigation of these concerns". Later evaluation, once the curriculum has settled and necessary initial adjustments have been made, will focus on changes in teaching standards and the attainment of pupils. The early monitoring and evaluation programmes have been strongly committed to working in partnership with teachers and schools.

REACTIONS TO THE NEW CURRICULUM AND THE PROGRAMME OF ASSESSMENT

Responses are bound to be contradictory - but this is the nature of all educational development within a divided society. (Simon, 1990)

The developments in the secondary school curriculum have been the subject of considerable support but also considerable criticism. The critics, perhaps inevitably, tend to be the more outspoken. The points that follow give some indication of the range of reaction.

- Uncertainty as to what the "basic" curriculum is, what "the whole" curriculum is and what features ultimately will have status in the eyes of pupils and parents.

- The overcrowding of the curriculum - ie the feeling that not much space is left for elements or emphases that reflect local conditions or strengths.

- Lack of evidence of real commitment to equality issues.

- The recognition that a curriculum which claims to be "an entitlement" curriculum for all is not in fact to be mandatory for all: pupils in private (ie independent) schools do not have to follow the national curriculum.

- The possibility that regular assessment against national criteria will prove to be an instrument of differentiation where some pupils could, early in their school careers, be labelled as failures.

- The extent to which the competition that is the hallmark of a free-market economy will devalue cooperation as a principle of procedure.

- The possibility that teachers as professionals may be, or feel themselves to be, deskilled and reduced, after a period of considerable professional autonomy, to being "agents" whose role is "to deliver" a nationally determined curriculum.

- 'The main test of the constitutionality of a national curriculum is whether it helps to promote personal and political self-determination' (White, 1988): there is doubt about whether the national curriculum supports commitment to such an aim.

- The absence of a convincing rationale for the selection of the core and foundation subjects. One critic has suggested that this is because the government cannot 'justify the political policy underlying the curriculum proposals', which he sees as 'squeezing out' social and political studies in order to ensure that 'pupils do not become critical of the society in which they live, (and) disposed to question authority in all its forms, not least in the workplace' (White, 1988).

Of course, it is easier at this stage to identify concerns than to produce evidence of good effects, but when it becomes possible to evaluate the effects of the national curriculum on teaching and learning, we must look beyond the subject-related results of the attainment tests to the issues of equity and self-determination raised above.

To offer a more positive comment, it is clear that in many schools teachers are working together - perhaps more closely than ever before - to examine the relevant contributions of their current practice, their aspirations for their pupils, and their teaching and learning strategies. There is also a new and proper concern for pupils' right to achieve. And some research (eg Ball and Bowe) is already indicating the ways in which the national curriculum as a "text" is being interpreted in different ways in different settings; the different interpretations testify to the possibility of both a committed intensification of the aims of the national curriculum and a conscious or unconscious subversion.

RESEARCH GENERATED BY THE NATIONAL CURRICULUM AND THE RELATED PROGRAMME OF ASSESSMENT

The national curriculum and the related developments outlined above have generated considerable research/development/evaluation activity.

First, there is the official development, monitoring and evaluation work (see above) which has been largely funded by the National Curriculum Council itself and by the School Examinations and Assessment Council. But, in addition, research has been funded from a variety of other sources. Here, I summarise briefly some of the main areas of this research activity.

- There is an intensification of research into progression in learning in different subjects.

- There is research into the coherence of the curriculum and the status and contribution of cross-curricular themes.

- There is research into the development and refinement of performance indicator systems; local authorities (to the extent that they maintain partnerships with schools) and individual schools are commissioning support in developing systems for interpreting examination results and for assessing the "value added" contribution of individual schools.

- There is research into the monitoring of the new inspection programmes; a particular focus is the link between inspection and school improvement.

- Recent research and development activities have focused on helping schools to construct appropriate "development plans" and to consider what function these might play within the framework of a national curriculum.

- Resources have been put into the development and evaluation of appraisal systems for teachers.

- Research is also focusing on the adequacy of the assessment system to capture the realities of pupil learning, the practicability of the assessment procedures - and the relationship between the formal attainment testing and the contribution of teachers' own judgments based on course work.

OTHER RESEARCH AND DEVELOPMENT RELATING TO THE SECONDARY SCHOOL CURRICULUM

- Youth employment and training: Since 1984 various Government Departments have funded a study of the routes taken by 16-year-old school leavers through training, employment and continuing education. (A parallel study has been conducted in Scotland). The data suggest the complexity of the routes, particularly during a period of national unemployment. Data also suggest that once young people leave full time education they do not easily find their way back. Other research focuses on specific aspects of the youth employment/training issue as well as young school leavers' views of what schooling has - or has not - offered them.

- Equality issues: A number of research and development initiatives have been undertaken on issues of equality (gender, race, disability and class). The funding in the main (but not exclusively) has come from agencies other than central government (whose attitude towards equality has been at best ambivalent and shaped largely by labour market demands). The focus of the various initiatives varies but the findings invariably indicate the need for sustained work in this area.

- Technology: A major developmental programme - the Technical and Vocational Educational Initiative (TVEI) - was launched by government agencies in the early 80s. It brought schools much needed financial resources for curriculum renewal, purchase of equipment, and for staff development. Technology continues, within the framework of the national curriculum, to be given priority, and industry - whose voice is increasingly influential in education planning - is underlining its importance.

- Motivation and engagement: Widespread concern about the underachievement of young people in secondary schools has generated a range of relevant research studies focusing on such things as discipline in schools, patterns of attendance and non-attendance, the identification and support of "at-risk" young people, and strategies for ensuring that young people are engaged in their learning and have an understanding of their own pattern of progress and achievement.

Postscript

I should like formally to thank all the people who provided me with information for this paper. While I am grateful for the help they gave me, it must be clear that the views that are expressed are my own.

SECONDARY EDUCATION AND
RESEARCH IN MALTA: AN OVERVIEW

Kenneth Wain
Dean, Faculty of Education, University of Malta

HISTORICAL BACKGROUND

Secondary education in Malta has gone through quite a hectic history of reform and counter-reform over the past twenty odd years to which this paper is limited.[1] Prior to 1970 secondary school provision was restricted to the successful candidates in a competitive 11+ style examination in English and mathematics to the State's grammar and technical schools, and to private schools (mainly Church schools). But in 1970, in the last year of the Nationalist (Christian Democrat) administration of the time, the decision was taken (following the reports and recommendations of two Unesco commissions, Lewis (1967), and Cameron (1970) to extend secondary schooling, and the first area secondary schools were set up alongside the grammar and technical schools.[2]

After 1971, on the other hand, with the advent of the new Labour government, radical changes occurred in the state secondary sector. The new educational philosophy of the Labour government for secondary education was an egalitarian one based on the principle of equality of opportunity for all. The government's interpretation of this principle led to the state grammar schools and technical schools being closed down and replaced by a common system of non-selective area secondary schools that aspired to operate on the lines of the British comprehensive schools. At the same time, in 1972, the Labour government also introduced another innovation into secondary education, the establishment of trade schooling which was intended to upgrade trade skills and produce new generations of skilled workers in line with the emerging economic needs of a newly independent country.[3] Meanwhile, the 1974 Education Act raised the school leaving age from 14 to 16 in two stages: Legal Notice number 3 raised the school-leaving age to 15 for the scholastic year 1973-74 and to 16 from September 1st 1974.

The former measure, the introduction of global, unselective, secondary schooling for all, was immediately controversial for different reasons, not the least because it included a reassessment of the current traditional academic evaluation procedures, playing down the importance of formal examinations and moving in the direction of non-formal cumulative testing. Other serious controversies arose over the policy of mixing students from different social backgrounds and with different academic competencies and dispositions unselectively. One popular complaint was with what was perceived as a serious problem of discipline and motivation in the new schools.

Another was the difficulty of coming to terms socially and culturally with this non-selectivity. A more minor controversy was over the abolishing of school uniforms and other marks of distinction between schools.

All these changes have to be assessed within the context of a schooling culture which had hitherto been very selective and traditional in its outlook in both primary and secondary sectors, and of a policy of implementation which utterly disregarded both these cultural constraints and the unpreparedness of teachers for such radical reform. Predictably, therefore, the reform encountered resistance from all quarters: parents, teachers, education officials and politicians. The subsequent years consequently marked a steep rise in the demand for private secondary education, which, unaffected by the reforms in the state sector, continued to operate as before on grammar school lines, and a concurrent loss of prestige and of the more able students on the part of the state schools. The outcome was the growth of private schools to a new prestige proportionately with the criticism levelled at the state schools. They became over-populated and inadequate as the pressure on them to accept more and more students grew from year to year. The state schools, on the other hand, may well have fallen victims to a self-fullfilling prophecy since the predicted problems with school discipline and poor standards were popularly regarded as having materialized as a result of this mobility of the better students towards the private schools.[4] Public discontent, in effect, probably fed also by teachers' obstructionism, grew to such a degree that the Labour government was constrained to reverse its egalitarian policy on secondary education.[5] In 1976 national examinations were reintroduced into the secondary schools, while in 1981 the government restored selective grammar schools which were called 'Junior Lyceums' (the choice of name 'lyceum' was an evident attempt to link-up with the 'pre-comprehensive' past when the boys' grammar school was known as the 'Lyceum').

This move, in turn, created a tripartite secondary school system made up of junior lyceums, attended by the more academically able pupils who passed a reintroduced 11+ type of examination in English, Maltese and mathematics, area secondary schools for the less academically able pupils, and the trade schools, which were ostensibly intended to train pupils who showed little interest in academic schooling and more of an inclination to take up some trade but which, in fact, became the destination of the poorer achievers and students classified either as 'unmotivated' for the purpose of academic study, once the best had been creamed off by the private schools and the junior lyceums, and the second-best by the area secondary schools.

In fact, the popular perception of the trade schools, both with the general public and, apparently, with the educational establishment, was as a kind of dumping ground for pupils with behaviour problems or with poor motivation to learn.[6] The change of government of 1987 saw no substantial change to this tripartite philosophy although there has been a clear attempt to introduce measures to liberalise it thus reducing its elitist profile, through a revision of the entrance exams to the junior lyceums and an extension of the subjects taken in the same exam intended to bring them within the reach of a larger number of pupils, and through an increase in the number of junior lyceum schools available to accommodate the increase. On the part of the trade schools, also, the stated policy is to reorientate the courses towards a broader educational programme which is not entirely vocational but includes also the humanistic and academic components of an all-round education, and to raise their prestige by attending to the qualifications and professional status of their teaching staff and by reorienting them towards their original vocational purpose. Finally, with the area secondary schools, the declared official policy has been to staff them with the same teachers and resources available to the junior lyceums and to permit mobility to the same junior lyceums for students who grow to the standard.

This trend towards liberalisation, however, as was observed earlier, has stopped short of dismantling the tripartite philosophy in any way, nor can it be regarded as the beginning of a procedure to move in that direction. It can only be interpreted as an attempt to render it less rigid and more humane. This is borne out by the publication of the National Minimum Curriculum which followed the publication of the Education Act (1988). This Act was intended by the incoming Nationalist administration of the time to provide a comprehensive law and statement of policy which would revise and update the provisions of education in all the different sectors, and was preceded by a white paper. Among other important measures, like the official conferment of professional status on teachers, for example, and the reform of the university, the new Act empowered the Minister of Education to publish a national curriculum and a code of teachers' ethics. As a result, over the next three years or so after the publication of the Act, the Minister published national minimum curricula for all the levels of the educational system from the kindergarten to the post-secondary.

The first National Minimum Curriculum for Secondary Schools in Malta's educational history was itself published in 1990 followed, not very long afterwards., in 1991, by a curriculum for post-secondary education. The National Curriculum which, naturally, reflects the government's educational philosophy confirms the tripartite secondary school system. Indeed, there currently appears to be a fundamental agreement between the two major political parties over this philosophy judging by the Labour Party's 1992 pre-election document on education.

POLICY AND ADMINISTRATIVE STRUCTURES

This brief and very rough account of the most significant developments in the history of secondary education in Malta over the past twenty years or so, has itself to be framed within the context of two important considerations. First, that there is no tradition for research and development in the history of Maltese educational reform. Second, that there is no tradition of broad-based and grass-roots consultation either. The implementation of the policies and the reforms for secondary education described above, even the momentous ones of the early 1970s, followed, universally, a top-down model with the major decisions being taken at ministerial level, and filtered through the Education Department to the schools and teachers, probably with internal consultation.

The Maltese educational system can, in fact, be accurately described as a centralized one both with regard to its administration and its policy-making machinery.[7] The incumbent minister of education and his ministry have always enjoyed extensive powers over the school system and these powers were confirmed and strengthened in the 1988 Education Act itself. The Department of Education is the agency through which the Ministry works. The schools themselves enjoy very little autonomy or powers to innovate and have virtually no control over the curriculum which is implemented globally through national, published, schemes of work, and assessed on national annual examinations. Curriculum planning and development are conducted at the Education Department, though the Department does not have its own particular identifiable research unit, though it does have a Planning and Statistics Section. The new policies and results are transmitted to the individual schools to ensure a certain standardization of teaching everywhere.

This notwithstanding the fact that school councils were set up, following the Education Act (1988), in 1989, for all the state schools in Malta. But the school councils have no real substantial control, or authority, over the curriculum at all. Although the published guidelines for their operation and administration invites them to make recommendations about curriculum matters to the teachers and the Department of Education, to advise about the school's physical and moral environment, and about discipline, and to be involved in promoting parents' participation, the administration of the school budget is projected as their major concern.[8]

SECONDARY SCHOOL REFORM

Apart from the trade schools and the prescriptions of the National Minimum Curriculum, the actual secondary school curriculum itself, its content and the style of teaching and learning involved is, as is to be expected in a heavily exam-oriented system like Malta's, strongly conditioned, particularly in the last two to three years of schooling, by the examination which lies waiting at the end and which, up to the present time of writing, has been the 'O' level examination set by the British Universities. This examination, however, together with the University of Malta's own matriculation examination system and the British style 'A' levels which it runs, is currently under radical review in line with a new policy intended to break off Malta's traditional dependence for secondary school certification on the British universities completely. Current government policy in this respect is to phase out the 'O' and 'A' level examinations which have, indeed, become grossly inadequate and irrelevant as well as expensive, and to replace them with a different local examination system which will be administered by the University of Malta in collaboration with the Education Department. The University of Malta is currently, in fact, at the time of writing, setting up the structures and the administrative machinery needed to take full control of the new local exams at both levels. And the indications are, though, again, at the time of writing there is no official statement on this score, that the new Secondary Education Certificate which will eventually replace the 'O' level will be managed on roughly the lines of the British GCSE with the A-levels being replaced by an examination of the International Baccalaureate type.

With regards to the curriculum itself, the most important reform undertaken recently through the Education Act, 1988, was, in fact, the institution of the National Minimum Curriculum which was, evidently, designed to redress the balance away from an existent rampant vocationalism towards a more broad-based education.[9] Another important innovation, however, was the introduction of a compulsory, post-secondary, course for University admission which was given the name Systems of Knowledge. The third crucial reform is the institution of the new local examination system referred to which, as was stated above, is currently still in its gestation phase though the likelihood is that the first syllabuses, and, therefore, the first official indication of the shape it will take, will have been published just before this workshop.

The National Minimum Curriculum,[10] as the name indicates, was designed to create a common set of principles for secondary schools in Malta and is binding on both State and private schooling. At the same time, its designation as minimum was intended to offer scope for an amount of flexibility in the selection of content. The document designates the aims of secondary education as being:

(a) the further development within widened dimensions of each student's intellectual, affective and physical abilities and potentialities;

(b) the training of the young mind in the pursuit of knowledge and reason, and the provision of a sufficient body of notions which would form each individual's basis for right judgment and proper value formation;

(c) the initiation into the process of qualification at a later stage for a working life.

The ambitions of the curriculum are also explicitly laid out in the document. The function of the curriculum, the document says, 'is not merely that of quickening the pace of each person's development and formation, but also that of moulding the nation, establishing a new cultural and technological profile for the whole country'. This is developed into two further objectives, one of them: (a) a workforce, in greater proportion literate and trained, qualified and specialized as well as able and flexible to retrain, respecialize or specialize later in a narrower field, expands directly on aim (c) above. The other, (b) a citizenry adhering to a wide ethical and cultural consensus on an increasingly higher level, should, logically, have been included as the fourth of the general aims stipulated by the curriculum. It is an intriguing question why it wasn't. The curriculum document also includes a section on methods: 'instead of being taught and instructed', it says, 'the student is trained in the process of self-learning and self-education and in becoming convinced of the importance of accepting (and not merely conforming to) norms of conduct. From activity and project work,' it continues, 'the students will be led to arrive at conclusions through research methods and problem solving.'

Finally, the learning programme itself is divided into 'core subjects' and 'options', the idea being that there should be a common curriculum for the first two or, preferably, three years of secondary schooling out of five, with options being introduced for the last two or three years. The core areas: (1) 'normative', including religious education, civics and environmental attitudes, and sport; (2) 'communicative', including Maltese, English and Mathematics as core, with a foreign language included in the first year and the possibility of another taken as an option in the third or fourth. For trade schools there is the further provision that technical English may be opted for instead of a second foreign language; (3) 'cognitive-scientific', including integrated science with a physics base, and geography, and with the options on offer being biology, physics and chemistry; (4) 'cultural accretion', including 'Maltese history within a wider context' and appreciation of literature, art and music, with the option being to pursue the more detailed study of the particular literature of a country or period; (5) 'action-work oriented', including life-skills, and technical design, with options being in home economics and particular commercial or trade choices.

Finally the document closes with a section on examinations which refers to the need to play down the negative effects of examinations by introducing more cumulative assessments, and refers also to the need to include an oral element in languages and in the core subjects, and a component of practical testing in the sciences and action-work oriented options. 'The greater reliance on oral testing', it contends in conclusion, 'should be aimed at preparing a more pronounced, oral communicative ability in our students'.

SYSTEMS OF KNOWLEDGE

Systems of Knowledge was a course which was introduced in a paper published by Paul Heywood and Peter Serracino Inglott towards the end of 1987 [11] and examined for the first time in 1989. 'If the School was to foster in our students a greater flexibility in adapting to changing patterns of work and life in a post-industrial age, it should afford them opportunities of going beyond the traditional limits of particular disciplines and gaining insight into different systems of knowledge'. (p. 181) This was an aim which, the authors argued, the 'A' level examination system was not achieving. The 'A' level examinations, they said, need to 'be supplemented by a cultural course which would not only help break down departmental separatism in schools, born of a concentration of effort on narrow 'A' level syllabuses, but also prompt students to reflect maturely on the specific learning derived from their 'A' level courses and help them to relate it to other fields of knowledge within a broad social and cultural framework.' (p.182)' What the examiners will be testing', they said, 'is the candidates' ability to grasp, and experiment with, ideas and principles and not simply the capacity for memorizing facts'. (p. 181) The examination would be demanding the same calibre of work as the 'A' level without being examined to that level.

The first curriculum for Systems of Knowledge identified the following areas of work and study:

Paper I (three hours)
Man and Symbols (sic.)
Man and Environment
Man and History: the Mediterranean and its role in the world.

Paper II (three hours)
Scientific Method and History of Science
Artistic Aims and Achievements

In 1991 a 'technological project' was also introduced into the course and the papers were re-designed to make space for this change but with the same knowledge areas included. There have been periodic meetings between examiners and teachers of Systems of Knowledge since the first examination, the first immediately after the first session in June 1989, but there is no published research to date on the course and changes within it have depended mainly on informal feedback and examiners' reports.

RESEARCH AND DOCUMENTATION

As was mentioned in the previous section, Maltese educational reform does not have a research tradition, nor has the Department of Education ever had a research branch or unit as such though it has always had a 'Planning and Statistics' unit and, very recently, it has appointed an officer in charge of curriculum development. It does have a Test Construction Unit which sets and assesses national examinations and compiles statistics, and does different kinds of data analysis. There is no national institute for educational research, however, and no publicly available documentation centre for education material either. Apart from the documents housed at the University Library, which are mainly restricted to official publications of statistics and reports and the occasional degree thesis, and the archives of the National Library, there is no documented material on education publicly available. Non-statistical qualitative research into Maltese education is, in fact, of very recent date and is mainly the fruit of work done in the Faculty of Education at the University of Malta either directly by its academic staff or through supervised B.Ed. and M.Ed. dissertations. Apart from this recentness of research initiatives, the very inadequacy of research facilities due to the unavailability or inaccessibility of documents makes certain research exceedingly difficult.

This research situation in Maltese education reflects an attitude towards research which, as was implied earlier in this paper, is reflected in the fact that none of the major reforms in education in Malta over time, including those outlined above, has been preceded by any officially commissioned research or subsequently monitored by commissioned research. Most have either been the outcome of foreign commissioned reports, like the Lewis and Cameron reports referred to earlier, or of the fiat of governments or high education officials. All the major changes in the system referred to in the first section of this paper were carried out without any published preliminary study or research or consultation document (the two Unesco commission reports mentioned earlier were not published), or any preliminary pilot study, and this includes, as was also mentioned earlier, the radical changes of policies and reversals of policies of 1970-1976 and the publication of the recent National Curriculum (1990). Nor has there been any systematic monitoring of these changes or innovations. In the latter case, publication was preceded only by select internal consultation within the Department and with some heads of schools, and an impact survey seems overdue.

Within this historical situation the price paid by society and by the individuals affected by experiments that somehow went wrong has, predictably, been high and far-ranging, but we are still substantially in the dark about their effects and some aspects of their politics although we are beginning to get to grips with them through the research of the small band of researchers at the University. The difficulties with conducting this research are considerable with the limited research time available for the researchers who are also actively involved in teaching different courses with large groups and in teacher training, with limited research assistance, and the limited availability of or access to documents and other data due to the lack of an efficient documentation centre referred to earlier.

The most immediate new challenge facing secondary education now is the change in the examination system referred to above. But, even here, there are no published reports or proposals preceding this exercise. Public, official, information about the projected reforms in the system, at least up to the point at which this paper is being written, is extremely scarce, notwithstanding the public's preoccupation about them.

What is known is mainly by indirect hearsay and through the very occasional unofficial remark or item of information heard here and there from persons involved in the exercise. It may well be that the reason for this state of affairs is that there is still no full-time, professionally trained body entrusted with the academic and administrative preparation of these examinations, and that the coordinators involved in the exercise are also engaged in many other things besides. But the stakes here are evidently very high because the failure of the new examination system would throw secondary education in Malta backwards by about 20 years.

It is evident, in the light of these reflections, that the time is long overdue for a drastic revision of administrative and structural policies in Malta, for the educational authorities to set up an effective, functioning public relations and information unit to keep the public informed of projected developments and changes in educational policies, and for these policies to function within a model of research and development which will ensure that they have a more solid empirical and theoretical basis than the opinions of officials and practitioners, however experienced, enlightened, or long-serving, within the educational establishment, and valuable as they may be.

Besides, and arguably most importantly, the current obscure and bureaucratic administration of education is an outdated relic of the old colonial system which has continued to survive virtually intact notwithstanding nearly thirty years of independence. The current revision of the administrative structures of the civil service as a whole being undertaken by the government, will, hopefully, infiltrate those of education also in the interest of greater democratic accountability and transparency, and, consequently, greater public confidence in the system at large, though the difficulties of moving in this direction cannot be underestimated.

RESEARCH AND THE FACULTY OF EDUCATION

The problem of persuading the people who count, the policy and decision makers, of these arguments can only be measured within the context of the negative traditions that have been referred to. The task is not an easy one though we should have experienced enough, on a national scale, of the negative consequences of unresearched and non-consulted reforms to make persuasion unnecessary. The other problem, supposing the persuasion works, is how we are to set ourselves up to respond to our research needs. Arguably the future of the Faculty of Education at the University of Malta is of crucial importance in this respect, although some recent unpublished studies undertaken through the initiative of the Education Department or its sections promises potential developments from those quarters also.[12]

In assessing this future at least two interconnected factors are especially important. One is the fact that the Faculty is a very young one. The other is a history embedded in teacher training. Indeed, notwithstanding some evident good progress in this direction, the Faculty as a whole has not, up to this stage in time, yet evolved sufficiently into a research institution. At the same time, its origin as a teacher training institution continues to militate against its public image with regard to research, and also, one suspects, its image with the Department of Education. Thus, the future of educational research in Malta, and hence of education itself, depends largely on its success in winning this image of a research institution and on a radical reassessment of its policy-making strategies by the Department of Education responding to the perception of the need for such strategies to be research directed and research monitored. For this to happen, the possible options, in theory, would seem to be either in the direction of a national institute and documentation centre with its own researchers set up in joint collaboration by the University and the Education Department, or for the Faculty of Education to become the Department's research unit in the sense that it is commissioned by the Department to do research in response to jointly-identified needs.

NOTES AND REFERENCES

1. See J. Zammit Mangion, (1988) *'An Analysis of the Expansion and Growth of Education in Malta since 1946'*, in C.J. Farrugia (ed.) *Education in Malta: a Look to the Future*. UNESCO.
2. See M. Darmanin, 'Malta's Teachers and Social Change', in M. Lawn (ed.) *The Politics of Teacher Unionism* (London, Croom Helm, 1985) and 'National Interests and Private Interests in Policy Making', *International Studies in Sociology of Education*, Vol. 1, 1991, for a socio-historical background and analysis of these reports.
3. R. Sultana has done extensive sociological research into vocational education in Malta since he launched his 'Trade Schools Research Project' in February 1989 with a special focus on the trade schools. The results of the project are due for publication as a book in late 1992 by Mireva Publications, Malta, titled *Education and National Development: Historical and Critical Perspectives on Vocational Schooling in Malta*.
4. See J. Zammit Mangion, op. cit. (1988). Zammit Mangion says that the 'headlong stampede to private secondry schools' had reached a stage whereby the 1980s the private secondary schools catered for 30% of the total secondary school population. (p.23)
5. See M. Darmanin *'Malta's Teachers and Social Change'* op. cit., 1985, for an account.
6. R. Sultana's study (op. cit.) provides a valuable sociological analysis of the history of the trade schools and of the social background of the students they attract.
7. See C.J. Farrugia's Introduction to *Education in Malta: a Look to the Future*, 1988, op. cit.
8. *Kunsilli Skolastici: Linji Gwida fuq ix-Xoghol u t-Tmexxija taghhom* (Guidelines for School Councils, undated official document).
9. For an in-depth analysis of the National Minimum Curriculum see K. Wain, *The Maltese National Curriculum: a critical evaluation* (Malta, Mireva Publications, 1991).
10. The National Minimum Curriculum (Secondary) was published through Legal Notice 103, 6th July, 1990.
11. P. Heywood and P. Serracino Inglott, 'Systems of Knowledge', *Hyphen*, Vol. V. No. 4, Msida, Malta, 1987.
12. A good example of research initiated by the Education Department, apart from the statistical data collected by its different sections, is the Tracer Study

(1990-91) on a national level of students completing their statutory secondary/ trade school course to see whether they had continued their education; if in the affirmative, where and what course, if in the negative whether they were in employment or registering for work. This study was carried out by the Department's Guidance and Counselling Services. The most recent Faculty research publication is R.G. Sultana (ed.) *Themes in Education*, Malta, Mireva Publications, 1991.

THE CONSORTIUM OF INSTITUTIONS FOR DEVELOPMENT AND RESEARCH IN EDUCATION IN EUROPE

Johan C. Van Bruggen

Secretary, CIDREE

INTRODUCTION

The 'Information Memorandum' for the workshop on 'Research into Secondary Curricula' formulates nine challenging questions concerning this topic. These questions involve the aims of education for the 11/12 - 18/19 group: the problems of reform, content, organization, and overload in the curriculum, and other important questions about the justification of the curriculum in the face of a changing economy, of growing information technology, cultural and political changes in Europe, etc.

It is impossible to cover all the issues packed into these questions in a short paper even if we take the situation in the Netherlands as our paradigm example of a country with a European outlook, based on its contacts with the Consortium of Institutions for Development and Research in Education in Europe, CIDREE. There is a lot of research relating to each of these issues going on. To begin with, there is the 'official' research documented in databases like ERIC, EUDISED and others, but there is also a lot of other research produced by university groups and by individual researchers in research institutions or in institutions for curriculum development.

CIDREE has become, since 1989, a consortium of 20 institutions for research and curriculum development in the western part of Europe. These institutions want to develop cooperation on problems in curriculum development and research: for example on projects about Values in Education, the curriculum for young children, adult education, and history as part of the core curriculum for 12-16-year-olds. CIDREE wants also to bring members of staff from participating institutions together to share information, experience and ideas.

Not only are there too many curriculum research areas to be covered in a short paper, but the research areas themselves are also distributed among hundreds of topics and issues. These vary, say, from investigations into the long term results of comprehensive education to, say, an evaluation study of the effects of five lessons in a certain part of the social geography curriculum emphasizing thinking skills.

In addition, there are very important and difficult language problems that limit access to research data. Many researchers in the curriculum areas only publish in their own languages, and much of their valuable research is only accessible to colleagues of their own nationality even if descriptors are fed into international databases. It is inevitable, for instance, that my own paper should have a Dutch-English-German bias.

In this paper I want to focus on three topics:

a. Research about a major curriculum reform in the last twenty years in many European countries: the creation of comprehensive schooling for all children aged 11/12 - 15/16. Although in many countries this reform originated from the concern about socio-economic inequalities and the hope to diminish the influence of these inequalities on the schooling careers of children, the creation of a comprehensive school was also a major curriculum reform in England and Wales, Scotland, Sweden, some parts of Germany, the Netherlands, Spain, Italy, France, and also in Flanders.
b. Connected with this is my second topic; the modernization of content and the delivery of content in the traditional subjects. In my view, the major changes that occur in the teaching of the traditional subjects are often overlooked. Thus, although it may be very clear to those who are near to the practice in schools that, for example, the teaching of geography is very different now from the way geography was taught in the 1960s, it may not be so clear to others more removed.
c. My third topic is the huge problem relating to the need to respond to claims on the curriculum for 12-16-year-olds originating from the debate on current social needs and problems engaged in by all sectors of society. These needs involve education for peace, drugs, Europe, a sound sex life, citizenship, consumerism, etc. Some questions in the information memorandum also raise this important issue. How can we reconcile the teaching of the traditional subjects, albeit modernized in their content and delivery, with the teaching of new subjects (information technology, technology as a general subject, economics as a general subject, a second or third language), and with new issues and topics (peace education etc.)? And how can we broaden the aims and objectives of education in preparing children for study and life in the 21st century in connection with the traditional subjects and the new subjects and topics? Also, the problem of overloading the curriculum is one of the most persistent problems in designing curricula for the 21st century. What can research tell us about the problems of curriculum developers and teachers?

The third topic is not dealt with in a separate sub-section but in the two sub-sections of section 3, where the first and second topic are elaborated. In the second section I elaborate and justify the statement that there is much valuable research in all kinds of topics and areas of the curriculum for junior secondary education but that huge problems exist in:

- the implementation of these research-based curricula for a part of a subject or a topic;

- bringing results and approaches together in a well-balanced and coordinated curriculum-as-a-whole, in which modern and research-based approaches for all subjects and topics are brought together;

- infusing ideas about broader aims (e.g. learning to think) and cross-curricular themes (e.g. environmental education) into a modernizing subject-structured curriculum.

I have worked as director for 'development' in the Dutch National Institute for Curriculum Development (SLO), so I shall, of course, take Dutch educational research in the curriculum for secondary schools as my starting point, although in the past seven years or so I have tried to complete my knowledge about curriculum questions in other countries, European and non-European. This interest has contributed to the origin of CIDREE, of which I am the secretary. In this paper, in fact, I shall try to connect the topics I discuss and my approach with examples from other European countries.

THE PROBLEM: IS A RESEARCH-BASED CURRICULUM ONLY USE-FUL IN LABORATORY SITUATIONS?

In the preparation stage of this paper, I saw by chance a description in EUDISED (9850) of a piece of research concerning music education by Kopinova, a researcher in the Educational Research Institute in Prague, Czechoslovakia. I quote the description:

> The aim of the study was to examine how the teaching/learning process in music education in primary and lower secondary school might be improved by enhancing the active character of the lessons. The study started from the assumption that varied, purposeful, musical activities will facilitate the accomplishment of the basic aims of instruction as laid down in the curriculum. The effectiveness of instruction can be increased by organizing musical activities into thematic units, accompanied by relevant musical material, and by placing emphasis on cross-curricular relationships. The research took the form of an experiment with specially designed instructional models in which pupils were observed. The approach appeared to be very effective in respect of the development of pupils' all-round musicality. The results indicate that it is necessary to elaborate music education curricula further in a way that allows for the consistent use of a creative approach to music. Special attention should be paid to tasks that are intended to develop the musical creativity of pupils. Teachers training should enable teachers to put the principles of active music teaching into practice consistently.

The description is illustrative of the problem of identifying the research base for curriculum work. It concerns a piece of research in a certain traditional subject, music. Its aim is to improve and update the content and method of music teaching and to connect music teaching with more general aims of education: to develop creativity, and to connect musical activities with other cross-curricular themes and issues.

In the research, musical activities are designed purposefully and probably based upon a fundamental analysis of musical phenomena. Important questions will be raised at this workshop that are relevant to this piece of research, about the aims of education (general and in a traditional subject), the content, and the problem of over-loading.

It is very easy to collect hundreds of reports of this type of curriculum research, not only in the Netherlands but also in most European countries. Such research reports show that a solution for a curriculum problem can be designed and elaborated into teaching/learning material and organization patterns in classrooms, and that the objectives formulated in advance are indeed met as a result of the relevant curriculum development.

Researchers have therefore proved that, for example:

- it is possible to design and do mathematics education in a highly successful way in mixed-ability classes in Dutch junior secondary schools through the implementation of a modern mathematics curriculum where high and low achievers learn to cooperate in small table groups (Herfs, 1991);

- it is possible to design, develop and deliver programmes for vocational preparation and initiation into the world of work for young people (14, 15, 16 years old) within both comprehensive and differentiated schools. (see: the Technical and Vocational Education Initiative in England and Wales; the Trawl Project in Northern Ireland ; similar projects in the Netherlands);

- it is possible to define objectives and create content for education about human rights and to infuse this content into more or less traditional curricula for history and geography, and to develop these curricula with good results (see, for example, materials released by the French Institut National de Recherche Pedagogique INRP in Paris);

- it is possible to develop models of series of lessons in political education and to deliver these lessons in controlled circumstances with good results (projects for social science in the Netherlands, in Germany and in other countries);

- it is possible to develop, evaluate and disseminate regionally a teaching approach aimed at stimulating discovery-learning through self-regulated activity for children in primary schools (in the Northern part of Germany for science, geography, etc., in the AKTIF-project (Hameyer, 1987);

- a last example: it is possible to develop a curriculum for the lower vocational schools in the Netherlands in which content from mechanical and electrical engineering is connected with content from traditional subjects like Dutch, English, mathematics, physics, through the thematic approach leading to the design and construction of windmills and bio-gas installations as examples of alternative energy resources.

What is the common factor in this type of curriculum design and research?

Researchers and/or developers try to demonstrate that a particular theory-based approach to learning and teaching can be realised in terms of curriculum materials and in the organization of the work in the classroom. The ideas may originate from new approaches in some particular subject, for example from an activity-based approach in science education in which the aim is for students to invent their own science linked to their everyday experiences, instead of acquiring facts, knowledge and behaviour through a teaching mode based on the teacher's demonstration. These ideas, internal to the teaching subjects, are often linked with well-known general educational aims, like learning to cooperate, learning to cope with new information, learning to learn, learning to assume responsibility for one's own learning process, etc.

There is enough good literature about the need to give more prominent attention to these kinds of general aims. I only refer to two small booklets. One, 'Curriculum Reform: an overview of trends', has been written by Malcolm Skillbeck and was published by the OECD in 1990 (Skillbeck; 1990). It is based on provisional reviews from various OECD countries about developments in the core curriculum for primary schools and junior secondary schools, and on the general curriculum literature. Skillbeck shows that it is not too difficult to formulate and elaborate these kinds of aims theoretically, but that, on the other hand, it is really difficult to realise them in practice in a consistent and comprehensive design of a core curriculum which enables one to see the pedagogy, content, organization and sequence clearly. Nevertheless, he also shows that there are many well-researched successful partial experiments and smaller-scale innovations in various countries.

The second small booklet is by Heinz Schirp (1987) from the Landesinstitut für Schule und Weiterbildung (LSW) in Soest, Germany. It is about the difficulties faced by the less able learners in junior secondary education. Schirp shows by describing examples of good teaching practice that it is possible to create more motivation in these learners, more ability to assume responsibility for their own learning, more positive socialization, more real cooperation between pupils and teachers and other persons and institutions outside schools, and so on. But it will be very difficult to expand these valuable approaches into a broad and massive innovation of the curriculum, because key factors in their success are the ability and willingness of teachers to change their pedagogy and to make professional links between the project-and-activity-based approach and the more mundane content of the subjects.

Both authors confirm the same conclusions arrived at in the other examples mentioned earlier: it is possible to turn ideas into a concrete series of lessons and activities; teachers are often involved in the design and preparation of material of this kind; teachers are usually trained for the purpose before some particular innovation is tried out, often with control groups. Evaluation shows the experiment to have been successful; students become active, enjoy the new approach etc. Hundreds of smaller and larger projects of curriculum development based on this approach and often carried out in cooperation with curriculum developers, teachers and researchers, have been reported in the literature. The curriculum itself is normally published in a well-presented book, sometimes also sold to a publisher, and recommendations are written about the relevant in-service training and the mode of implementation.

This is where the problems begin because large-scale implementation is difficult. Much of this type of innovation only reaches a small number of teachers, often those very same ones involved in the experiment and the development project to begin with. And even these often find themselves having to water down the innovation in their daily practice over the years. This is the situation described by Goodlad (1984) in a large-scale study called 'A Place Called School'. Goodlad's research showed that very few of the results of the hundreds of projects in curriculum development and research from the 1960s and 1970s in areas like science teaching, mathematics teaching, social science teaching, arts, etc., could be detected in the everyday practice of the average American high school. An investigation similar to Goodlad's has been carried out by Kuiper and Alting (1990) into science education in the Dutch junior secondary schools. Although some of the results of various projects for biology, physics, chemistry, and environmental education can be found in textbooks and in the average practice of teachers, the results are poor compared with the massive investment which is made in curriculum development, evaluation, research, etc. There is no reason to believe that in other subjects or areas of the curriculum the situation is better.

Recent reviews of the curriculum situation in American high schools, written by Cohen and Spillane (1992) and Gehrke, Knapp and Sirotnik (1992) published in the eighteenth Review of Research in Education (Grant, ed., 1992) show the same findings. Cohen and Spillane in particular are very pessimistic about the possibilities for a major reform in the curriculum. Their explanation for this pessimism is that the educational system in the United States - which is very decentralized - makes it impossible to come to a coordinated national reform that would infiltrate hundreds of thousands of independent classrooms. Gehrke, Knapp and Sirotnik review the research about what schools are really teaching. Regarding social studies, language, art, mathematics and science the situation is that, although these demonstrate gradual changes and improvements (in particular in the teaching of mathematics and science), most of the teachers still deliver a poor curriculum which reflects very few of the principles of curriculum development and research.

What is the problem with the secondary school curriculum?

The problem is not that we lack ideas at the level of recommendations about the curriculum (Glatthorn, 1987); about aims, objectives, and content. All official curricula released by government or official agencies have fine chapters about general aims of education, about humanistic education, about respect for international solidarity, respect for other communities, democratic behaviour, etc. The problem is not that we do not know how to design examples of different aspects of the curriculum in which these aims are elaborated into materials and guidelines for teaching and learning behaviour and classroom organization, either. There are very fine examples of curriculum materials produced through different projects all over the world, including Europe. Neither is it the problem that these fine examples of good, modern, balanced, integrated aspects of the curriculum are insufficiently researched (See for masses of information that backs up the validity of curriculum research the International Encyclopaedia of Curriculum (Lewy, 1991), for example). Nor is it the problem that we cannot demonstrate the success of these innovations in small scale, controlled, circumstances, where there has been ample investment in researchers, curriculum developers, teacher trainers, teacher time, materials, etc., although the innovations have proved difficult to sustain for longer periods (a whole school year, the course of several years) without the continuing support of the teachers involved. The serious problem is that, in spite of all this development and research, the 'taught curriculum' (Glatthorn, 1987) in the average classroom in Western European countries and in the United States of America does not reflect the fine 'rhetoric' about the curriculum , and the small-scale successes disseminate only very gradually and very piecemeal. The research into innovation and implementation in education (for example, Lieberman, 1986 and many others) bears this fact out.

But my view is that the problem is more than one of implementation as such. I have already mentioned the difficulty of designing a curriculum at the operational level; a curriculum that plans the whole vast tapestry of thousands of lessons and activities in which the best ideas are connected, coordinated, and balanced. Such a 'weaving process', in fact, involves attention to numerous factors. This is especially so when what is involved is the longitudinal planning of concept-building and skill-training in separate or connected subjects, but it is also the case when the task is to plan how general aims can be served by curricular activities in general subjects. The point is very well illustrated in the research and development of 'learning how to think' (as a general aim). The report of the major OECD-conference of 1989 (see Maclure, 1991) has good theoretical chapters about the infusion of 'thinking' in the general curriculum, but few writers suggest how this can be done concretely on a large and systematic scale.

The problem of design becomes even more complicated if we take into account the many claims made on the curriculum to accommodate new topics like peace education, environmental education, and so on. CIDREE organizes a regular workshop (this month it will take place in Belfast) about curriculum development in these so-called cross-curricular themes. In my own institute, SLO, we have counted 60 of these cross-curricular themes. The problems involved in identifying core-objectives for each of them, developing well-focused curricular activities etc., can be described with the same remark about the difficulties of implementation on a larger scale. The problems of integrating these claims in a balanced and well-organized curriculum are gigantic ones involving not only the factor of a very complicated curriculum design but curriculum overload also. They are very well demonstrated in a UNESCO book which, in my view, has met with too little attention.

The book, published in 1987 and entitled 'The Contents of Education: a worldwide view of their development from the present to the year 2000', was edited by Rassekh and Vaideanu with the participation of authors from various countries, and based on a long-term UNESCO project. It contains analyses of important developments and forecasts about trends and issues involving topics like population growth, socio-political changes in the world, economic changes, technological developments, etc. These trends and issues are seen as possible sources for the content of education during the time ahead. The second part of the book sketches out suggestions about new content for education programmes. It is the best example, to my knowledge, of a review of the 'rhetoric' about issues like education for citizenship, technological education, moral education, productive work in education, interdisciplinarity, a new understanding of the world, active thinking, a creative new understanding of social phenomena, etc. Rassekh and Vaideanu take up the challenge of designing a new syllabus in which all these demands are mixed with the traditional content of the subjects. But no sharp choices are made. Solutions to the overload in the curriculum, they argue, should be found in interdisciplinarity, project-based work, active learning etc. In the conclusive part of the book Rassekh and Vaideanu recognize that it is almost impossible to develop a complete list of desirable curriculum content because it inevitably grows in the process and this makes it very difficult to organize in a multi-dimensional, balanced and integrated way without dropping important traditional content from history, language, and mathematics, for example. They list five possible solutions to the problem, however:
- the promotion of interdisciplinarity in the organization of content and in the design of the teaching/learning process;
- the principle of lifelong education;
- the contribution of informatics and computer-assisted learning;
- the assistance provided to the school by parallel education, and, in particular, the mass media;
- the 'infusional' approach for the promotion of new types of education (eg peace education) in the teaching/learning process.

Two of these five solutions try to co-opt other agents outside the school for educational tasks and this - if it can be made to succeed in some countries, and therefore shown to be feasible - will enlarge coordination problems between school learning and parallel learning. Although we know that children learn a lot from outside the school by means of incidental, informal or parallel learning, it is difficult for schools to harness this learning systematically in planning and designing the teaching/learning process. Apart from the fact that its effects vary with different children, teachers do not have the instruments to assess this incidental learning, so planning is very difficult. An interesting investigation was reported by the Dutch researchers Streumer et al. (1987) into the need for introducing a new subject called 'technology education' into the junior secondary schools in the Netherlands. The objectives identified by, and the content created in, the curriculum development project of the SLO were used in a battery of tests. It appeared that students of 16/17 who were in the more advanced tracks of the differentiated Dutch school system mastered these objectives without any formal teaching, while students from the lower tracks in general education - and, therefore, with lower capabilities - did not master the same objectives, perhaps because these students are less active in reading, watching television, picking up information in informal settings, etc. So, using other agencies outside schools is very difficult to plan.

Regarding interdisciplinarity and the 'infusional' approach, we know from research that it is possible to work in this way. But the problem referred to earlier also appears here; it is certainly possible to adopt these approaches on a small scale under controlled circumstances but very difficult to realize in large-scale innovations. Rassekh and Vaideanu recognize these difficulties. Their explanation is that there is a growing national resistance to change on the part of education systems. Over and above the financial and technical problems and the divergences in ideas and interests about the content of education, this resistance to change is very important (see also the review of Cohen and Sillane and of Gehrke, Knapp and Sirotnik mentioned earlier). Rassekh and Vaideanu state that it has grown in the last 30 years through the expansion of the bureaucratic and administrative side of education, more rational management, more standardisation of procedures and assessment, and less opportunity for teachers to be engaged in genuine participation, as also through the failure of important qualitative reforms in many countries and through the diminishing financial priority given to education since the recession of 1973.

The last solution proposed by Rassekh and Vaideanu, the contribution of information technology seems not to be very realistic, as many failed projects have shown. The problems of overload and integration here are too complicated for it to work. So, in sum, although Rassekh and Vaideanu's book is an intelligent and rich analysis of claims on the secondary school curriculum, the solutions it offers to the main problems identified are of dubious worth. These problems we can now summarize as follows:

1. It appears to be very difficult to create a consistent and complete design for a curriculum to a detailed level of specification; a design in which the results of the research-based development of the different elements of such a curriculum are

brought together in a balanced, integrated, and transparent way which serves the stated general aims that are formulated for it.

2. It is difficult to design a curriculum which combines different researched-based proposals for new approaches to the teaching of the more traditional and the newer subjects with each other and with new topics, and with so-called cross-curricular themes, in a balanced, integrated, transparent way.

3. Even research-based proposals for innovation for parts of the curriculum-as-a-whole are difficult to implement on a larger scale. The actual taught curriculum only changes very slowly and with much watering-down of the original design.

These problems are also touched upon in a very recent article about the state of the art in German curriculum research over the last 10 years written by Uwe Hameyer (1992), who was involved for a long time in some larger research-and-development projects involving science and information technology programmes. Hameyer is not optimistic. There are, he says, very few larger projects and, in any case, with some exceptions in the early 1980s, these have not been based on extensive and sustained research. There is little conceptual analysis and few links with the literature, and, in the case of language, consultation with projects involving other language areas is rare. There is little continuity in staff cooperation, money and institutional support is weak, and there is too much investment in small, superficial, and short-term projects. He mentions six main research topics that have received attention:
- the topic 'essentials in the curriculum'; looking for criteria to make curricular choices;
- the problem of reconciling 'old' subjects with the 'new' and with new cross-curricular themes;
- the function of prescribed curricula and their implementation;
- the gradual changes in the processes of development involving the prescribed curricula and the growing link of these processes with participatory models and loops in implementation;
- the topics of 'activity' and 'discovery' learning in particular primary schools;
- the development of more complex teaching/learning arrangements.

It will be clear that the three key problems I mentioned earlier are reflected in Hameyer's analysis. Although he does not mention explicitly the problem of integrating all kinds of results and ideas into one larger curriculum design, some of these themes touch on this overarching problem.

In the Dutch National Institute for Curriculum Development (SLO) various projects for curriculum development, partly based on research about content and on changing demands and partly on cooperation with researchers, were carried out over the period between 1976 and 1991. Many of these research projects have investigated the impact of curriculum research on the educational system, on teaching behaviour in the classroom, on textbooks, on examinations, on the behaviour of the government.

A review of some 40 of these research reports about the impact of SLO projects (see Van Bruggen, 1992) shows that many of the projects had some impact on textbooks and on examinations or lists of attainment targets over the past few years but their immediate impact on classrooms and teaching behaviour is much less. Some larger evaluation studies carried out in a sample of Dutch schools show that the impact varies from 3% (dance in primary education) to 40% or 50%, where 'impact' is defined as observable effect on teaching behaviour (content, organization of the teaching/learning process, teachers' activities). In, for example, the evaluation studies about the impact of the projects on civics teaching (Kuiper c.s., 1989), and on science education in primary schools (Van den Akker, 1988), and in many others, it appears that the more difficult and progressive changes proposed by curriculum developers based on research into learning theory and into smaller scale projects, are very difficult for teachers to realize in their everyday practice, because they have insufficient time for preparation and organization, lack didactical skills, are uncertain about classroom behaviour, etc.

Research about the impact of other curriculum development institutes in various countries show the same slow or poor implementation if not worse. (Van Bruggen; 1987).

All these facts about the slow change of practice in education that are brought to light by research and analysis point to some common conclusions: it is possible to develop progressive and sound ideas about the renewal of the curriculum, it is also possible to elaborate these ideas in operational teaching behaviour and in new materials, etc., and it is also possible to demonstrate by research and by evaluation that these new ideas are successful in controlled circumstances and on a reduced scale, but it is very difficult to bring out the same improvement on a larger scale, and it is very difficult also to combine the results of projects for aspects of the curriculum into a design for the curriculum-as-a-whole.

Does this mean that it is foolish to invest in educational research in particular areas of the curriculum; in curriculum development and in in-service training and implementation, because it may never be possible to extend the intensive approach in developments on a reduced scale to the broader scale of the whole educational system? The next section will show that this conclusion is too pessimistic.

TWO MASSIVE FORCES PROMOTING CURRICULUM CHANGES: 'GOING COMPREHENSIVE' AND 'MODERNIZATION'

In the previous chapter I have made it clear that it is difficult to obtain curriculum change unless the implementation is on an experimental basis. Does this mean that important and research based curricular changes are almost impossible? Research on implementation processes (cfr. Van den Berg, Hameyer & Stokking 1989) shows that it is possible only if there is adequate coordination between the various sectors of education including the managerial level, school-based organization and researchers. This exercise could prove to be very difficult and expensive especially in democratic countries (cfr. Van Bruggen 1989).

Teachers are normally in contact with university lecturers through in-service training. They also have contacts with their colleagues, attend conferences, look for new insights into educational methods and contents. This is a more or less autonomous process in the renewal of educational content and in new approaches in teaching. It is in many ways a process of updating one's own resources. And several factors could contribute to this free and autonomous way of upgrading and updating one's teaching profession, namely a good strategy for in-service training and cooperation between researchers, curriculum designers, university professors, education publishing houses, teacher training institutions, etc... All these factors or/and change agents together with projects and planned research and training programmes could truly bring about an ongoing change in education.

This holds for the renewal of traditional subjects such as chemistry, modern languages, etc, especially if we take into account a longer period of some 15 years or more. But it is also true for the gradual introduction of parts of the so-called interdisciplinary subjects. Although I must admit that I am not *au courant* with many studies from European countries, I have come across a number of studies which have given me a general indication of the state of the art similar to Goodlad's study of 1984. But we know something from the various studies of the IEA about science and mathematics. From other studies in various countries, of course, a more systematic analysis of textbooks, examinations and results of some national-assessment schemes (United Kingdom, Sweden, France, the Netherlands) shows that things are not the same as they were 20 years ago ...

A second approach is that of 'going comprehensive' as with almost all western European countries. In a number of countries a type of comprehensive school for 12-15/16 has been introduced, as in the case of Sweden, France, Italy and in most areas in England and Wales. But also in countries like Germany, the Netherlands and Austria more and more students are choosing the higher tracks of general education (Gymnasium, Lycée ...) and this leads to a situation wherein schools can no longer offer a curriculum for an elite group (10-15% of the school population), but have to adapt their curriculum and their methodological procedures to masses of 50-80% of the age population. This cannot be done without important changes in the programmes of studies and in pedagogy. The result is that in some countries the Gymnasium is becoming a sort of comprehensive school for all children. This process has undoubtedly brought more curriculum change than was envisaged.

Comprehensive education as a compromise: a new impulse for curriculum changes

In most countries the setting up of one comprehensive school for 11/12-15/16 instead of schools for level-differentiated general education and for vocational education

67

stemmed from the political debate about equal opportunities for children from different socio-economic levels in society. But in all processes in Sweden, Germany (von Henting 1971), England (Hargreaves, 1982), the Netherlands, and also in France and Spain, the curriculum was also at the 'mercy' of experiments , debates, reforms and counter-reforms. The curriculum is the expression of the conflicting views on the cultural development of young children, of education in its most important aspects, of a participative society in developing democratic values and norms. In 1991 Roger Standaert published a fascinating comparison of the rationality in the policies for junior secondary education in France, England and Wales and Germany (West). His book entitled in Dutch (Flemish): 'De vlag in de top' (The banner at the top) indicates that a lot of rhetoric about comprehensive education and equal curricula for all children, etc, always remained at the level of discussions and debates between politicians, researchers, educational scientists and lobbyists and failed to filter into the real practice of schools. Standaert compares the results of all these changes in the three countries mentioned in three important curriculum areas and asks:

- is the separation of traditional subjects replaced by a more integrative, balanced and interdisciplinary curriculum?
- did systems of education succeed in emphasizing applied knowledge instead of abstract scientific knowledge and in linking school education with the reality of children in their homes and in the street?
- did the systems of education succeed in creating a broader education for children (emotional, physical, moral, political, social ...) instead of predominantly intellectual education?

Standaert (1991) shows, on the basis of evaluation of the results of the reforms, that changes did come through but not as dramatically as the godfathers of the systems had predicted and wanted. In England, France and even in the few German comprehensive schools separation of traditional subjects is common, although in some of the laws and regulations 'thémes transversaux' or 'inter-disciplinary themes' are compulsory and students are tested through formal examinations. As mentioned earlier, a lot of work has been invested in the further development of this approach in many curriculum development projects in the last twenty years.

The same goes for linking school education to applied knowledge and reality. One may take a closer look at the attainment targets, for example, in English education since the Education Act of 1988 and the attainment targets for the new compromise for Dutch basic education for 12-15/16 (see Van den Brink, 1992); bearing on this topic are also recent papers for Lycée and Collège from the National Council for the Programmes of Study in France (Ministère de l'Education, 1991 a and b), the Libero Blanco for the impressive Spanish innovation (Ministerio de Educacion y Ciencia, 1990 a and b) and the new curricula for the Gymnasium in Bavaria (Bayerische Staatsministerien, 1990) and the Gymnasium and Realschule in Baden Württemberg (Ministerium für Kultus und Sport, 1984). From such examples a number of attainment targets and regulations can be observed with a better balance between scientific, abstract knowledge and skills in practical situations. From research, for example, in the IEA projects for science and mathematics (see Postlethwaite and Wiley, 1991 and Rosier and Keeves, 1991 for the more recent science results) we know that some of these strategies are being tried out in schools. We already mentioned reports on science education in the Netherlands. Kuiper and Alting (1990) show the same gradual improvement in science education as a whole. In one of the major projects in the Netherlands - the PLON (1973-1985) for physics in secondary schools - this coupling of science and real life was one of the main goals. The interesting review and evaluation of Wierstra (1990) shows that development and elaboration of this principle was not easy, but possible. It was partially developed on the basis of research on : patterns of interest of children of various ages (12-18) and sex (an important research issue in science and maths development); their real life experiences in sport-home-disco-transport etc.; learning concepts by analysis of experience and simulation, etc. A lot of development with experimental classes and teachers has been combined with good formative evaluation and initiatives for dissemination. New textbooks, new examination programmes, new laboratory materials, in-service training, regional conferences, a strong link with the teachers' associations ... have been instrumental in the implementation, on a large scale of this project(estimated real impact 30 to 75% in various school types); but in most cases it was not as revolutionary as was expected at the beginning of the project.

More difficult is the situation with broader education as a third characteristic. Although there is enough rhetoric in laws and lists of attainment targets, we do not know enough about the practical realisation of these aspects of education. The research of Standaert, supported by the well-known research of Fend (1982), Aurin (1986), Hargreaves (1982) and the study of Goodlad about American junior secondary education, shows that it is difficult to illustrate that the comprehensive school is a necessary tool for bringing about more social equality through education. But the opposite is not true either, and research results increasingly show that in a moderate system of comprehensive education the bright students are not held back in their progress. The less bright students are also stimulated. [See also the research of S. Marklund (1980-1984) and other research from Sweden, for example, Ball and Larsson (1989)].

More important is the social and cultural setting that comprehensive schools of a moderate type provide. This setting makes it possible to broaden the curriculum for all children and to infuse richer educational objectives and new issues.

This is exactly what is happening in the Netherlands. After a debate of more than 20 years, various small-scale experiments, and the design of two new curricula for comprehensive education by the Dutch SLO (1982, 1985), government decided to consult the Scientific Council for Government Policy about secondary education. The council published its views in 1986 (Wetenschappelijke Raad, 1986). To my knowledge the book contains one of the best reviews and analyses of research about comprehensive education. The Council stressed that a structural change of the four school types for Dutch children 12-15/16 years old into one comprehensive school was impossible for political reasons and that it was not demonstrated that such an important change was necessary to meet the goals of educational policy i.e. personal and social development of all children, cultural and social preparation for life in a modern Dutch society and preparation for work and further study.

The Council therefore recommended the introduction of a broad and balanced curriculum for all children. During 80% of the time, every pupil would be offered the same in terms of subjects and attainment targets. This would afford both schools and pupils room for their own subject choices. It would also enable schools to retain some control on the pace of pupil progress and help ensure that the examination system is in line with that of further education and training. In this way it would be possible to have the content of 'comprehensive education', as it has been gradually developed also in some other countries, 'infused' into a system of separated school types.

In another Flemish cooperative study by Wielemans and Vermeerbergen (1990) as well as in Standaert's work (1991) this Dutch experience is described and compared with educational practice in France, Germany and England and Wales.

After protracted political wrangling, in the summer of 1992, government and parliament agreed on the core of the Council for Government's policy proposal, which will now be introduced in 1993. The curriculum prescribed is a package deal between conflicting interests and lobby groups for certain subjects or topics. It is born of power conflicts between the central government and the free schools and associations, apart from other sources of social tension.

What is remarkable is that this package deal is very similar to those put together in recent years in England and Wales (for a recent review of developments and problems vide Coulby and Bash (1991) or Flude and Hammer (1990), Spain (Vide the English translation of the Libero Blanco published by the Ministerio de Education y Ciencia, 1990b) and France, where the work carried out by the Bourdieu/Gros committee (1989) has culminated in the recent work of the National Council for Programmes of Study in Paris (Vide Ministère de l'Education Nationale, 1991a, 1991b). One could also mention the latest developments in California, Texas, Illinois and other American states.

What are the essentials of this new prescribed curriculum for a modified type of comprehensive education?

- Maintenance of the traditional subjects and their structuring effects on qualifications of teachers, timetables, examination systems, planning in-service training etc.; but using the labels of subjects for gradual renewal of the content. Of course it is very important which 13, 14 or 15 subjects are regulated in this package deal. We see - not only in the Netherlands but also in other countries - that ultimately lobby groups, members of parliament and people close to ministers have an influence far greater than that of scientific analysis or curriculum councils.

- Recognition that there are a lot of issues and topics, such as peace education, the European dimension, learning to think, environmental education etc., that deserve attention, and in formulating regulations outlining the attainment targets for more or less traditional subjects one must cover these themes, and measure student perform-ance therein, through examinations or national assessments. Prescribing through national regulations the infusion of such themes in traditional subjects means that they are no longer covered on a voluntary basis. This is the case in England and Wales, France, Spain, Baden Württemberg, Bavaria and, starting from 1993, also in the Netherlands.

- Formulation of attainment targets to a degree of specification that affords the possibility of guaranteeing some vertical coordination between primary, junior, secondary and upper secondary education and at the same time some horizontal co-ordination at national level between traditional subjects and cross-curricular themes in junior secondary education. However such detailed formulation of attainment targets should not rule out all freedom for curriculum developers and schools. This, of course, is a very important concern of research into the processes, formulation systems (Vide Van den Brink, 1992) and political forces begetting the package deal (Vide Van Bruggen 1986).

In all the countries so far mentioned we notice the adoption of attainment targets of this nature. Furthermore, in England and Wales, France and the German Laender of Bavaria and Baden Württemberg, it is also programmes of study that are prescribed. This is not the case in the Netherlands owing to the traditional freedom of curricula for free schools.

- Allowing enough room, in the formulation of these attainment targets for schools to exercise their professional responsibility in adapting to their particular social, cultural and economic environment and in catering for the needs of their students. In the Netherlands, for example, this principle is safeguarded by allocating 25% of the curriculum time in the three-year period of basic education between 12 and 15 to free subjects, to be chosen by the schools. In England the percentage of time allocated to these optional subjects is the same, while in Spain it grows from about 20% to 50% over the four-year period of the middle school.

In the Netherlands, these elective subjects could include religion, vocational preparation, Latin and Greek for potential Gymnasium students, extra mathematics, photography, philosophy etc.

Another principle adopted in this connection is establishing some limit to the curriculum time allowed for the attainment targets. In the Netherlands such bounds are tabulated for schools in an official document that is merely of an advisory nature; only 75% of curriculum time should be used for realizing the prescribed attainment targets. The rest of the time may be used by schools for interpreting the attainment targets, highlighting special applications, etc.

- Investment in the development of teaching aids and in in-service training for teachers and school managers for the realisation of this new package deal in the curriculum. This activity involves working out and sifting examples of programmes of study, examples of interpretations of attainment targets, examples of the infusion approach for cross-curricular issues, examples of assessment schemes for attainment targets, examples of classroom organization, examples of horizontal coordination etc. In most countries this type of curricula production is organized by national institutions or councils, sometimes - but alas not always - in coordination with supporting research of various types.

So we see that thirty years of educational battles over comprehensive education in western Europe have now led to most countries adopting a moderate system of comprehensive education, with variations. But perhaps more important is the fact that a new, by no means revolutionary, package deal for the curriculum in junior secondary education has been developed. This is not a total restructuring of curriculum thinking on the lines contemplated by Robinson (1967), Von Hentig (1971), Lawton (1975, 1982) or Skillbeck (1990).

In my opinion, however, it affords enough room for improvement allowing as it does for the gradual introduction of new elements stemming from sound curriculum research and curriculum development. It is the autonomous system of implementation and professional renewal in the teaching profession that makes this possible. I have mentioned a lot of examples already; in all countries it is easy to discover these examples. International exchange and cooperation in the production and elaboration of new and sound practice within this package deal should be very fruitful and could be promoted by organizations like the Council of Europe, the OECD and the European Community; CIDREE tries to do this, but on a low budget scale. See for example Van den Brink and Hooghoff (eds.) 1990.

It is true that in the western part of Germany comprehensive schools as such have not developed on a large scale, but also in Germany the autonomous development into a general school for all youngsters is very strong and this has led to more and more youngsters opting for the Gymnasium; then for the Realschule and, only if necessary, for the Hauptschule. See also Schirp (1987). A comparison of the curricula prescribed by the ministries of Baden Württemberg and Bavaria is very interesting. In Baden Württemberg these new curricula for secondary schools of the different types were published in 1984, whereas in Bavaria in 1991. One could say that these curricula have exactly the same characteristics as the package deals in England, the Netherlands, France and Spain (see above). A very interesting paper (Staatsinstitut ISB 1990, not yet published) of the Staatsinstitut für Schulpädagogik und Bildungsforschung ISB in Munich explains why also in Bavaria, where there is no comprehensive system, the same is none the less valid for the curriculum of the

Gymnasium. Horizontal and vertical coordination has become essential, because there are so many claims made on the curriculum, also in the Gymnasium. Efficiency has to be improved. A huge operation of revision of the curriculum took place in the years 1986-1990, accompanied by a lot of research on desires and problems of teachers, parents, employers, churches and so on. The result is a curriculum that has been updated within the subject structure, with much more prominence given to cross-curricular issues and with enough room for schools to make their own choices.

In most European countries these curricular package deals are not the result of structural, scientifically based new designs of the total curriculum for secondary education, as we saw in the 1960s and 1970s. However the work carried out by experts like Von Hentig, Legrand, Lawton and bodies like the Dutch Innovation Committee for Comprehensive Education, the Dutch National Institute for Curriculum Development SLO and many others has had an influence on the formulation of the package deals. The package deal, in fact, is the result of a deft balance between various tensions resulting from political forces, the limited capacity of the educational system for radical change, far reaching educational perspectives and many small but significant shifts in thinking about certain areas of the curriculum.

In the years to come curriculum research could play an important role in scientifically laying the foundation for further curriculum development within the package deal. There are still many important areas calling for research and development. The following are some examples:

- What knowledge and skills are young children learning in the educational environment outside schools: by watching television, through sport, travel, popular music and so on? How can formal education in schools make use of this incidental and parallel learning?

- What type of knowledge, insight and comprehension is needed for children at the age of 14/15 to make responsible choices in optional courses and patterns of further study and vocational preparation ? An analysis of decision patterns and careers of students between 14 and 20 can be very helpful for defining the type of knowledge, skills and insights required for enriching the curricula in the subjects listed in the package deal;

- What type of skills in independent study, thinking, planning, self-control etc. are essential for a good start to further education after 14/15? What are the best didactical procedures for infusing these skills into traditional subjects? What type of didactical arrangements and organization in classrooms and schools give good results?

- What type of cultural knowledge and appreciation is essential for young children of 15 and older in order to profit from the new possibilities of a more open Europe, not only in the economic sense, but also in the cultural sense? Knowledge about languages, travel, history, geography, political systems, everyday life, etc. How can this knowledge be infused into the lists of attainment targets and into textbooks? What is learned incidentally? Couldn't we start a project to define 'cultural literacy' (cf. Hirsch, 1987) for European citizens?

- What is - in specific domains of the curriculum - the best mixture of expository teaching, homework and active exploration of reality? General theories about learning and instruction must lead to practical applications at the chalk face.

- What do young children know about computers, their use, their potential, their restrictions? How do they pick up incidental learning in this area? How can teachers and textbooks profit from this autonomous development?

This type of research is not research for a new grand design, but for patient and down-to-earth work involving cooperation between teachers, researchers and curriculum developers in order to support the educational system with instruments and materials that can gradually help to improve the quality of the taught curriculum.

Renewal of the curriculum in the 'traditional' subjects

In Europe we don't have a systematic and detailed study of the real curricular situation in secondary schools comparable to that of Goodlad (1984). Nevertheless, there is a lot of information, based for example on analyses of results of examinations. In the Netherlands the National Institute for Educational Measurement CITO is charged with the preparation and organization of the annual examinations in secondary education and each year an analysis of results is presented. Also in some other countries, this type of information is at hand. There is also information on some subjects in some school types, not at system level and not too detailed, but significant nevertheless. I already mentioned Kuiper's and Alting's (1990) investigation of the situation in science teaching in the Netherlands. A similar study was carried out by Bouwens and Oud-de Glas (1991) on the teaching of foreign languages in junior secondary education and another by Stoks (1989) on the teaching of foreign languages in upper secondary education. Some work has also been done for economics, history, geography, Dutch, technology, information and technology (see Van Weering and Plomp, 1991).

From an international comparative standpoint the project 'The curriculum redefined' started by OECD in 1987 was promising. One of the first results was the sample of small country studies, in which facts and figures about the curriculum, as it was prescribed and delivered, were reported. Skillbeck (1990) wrote an interesting review of the curricular situation in OECD countries, but on a rather general level and without detailed information about objectives, content, didactical arrangements and results. As far as I am aware, detailed country studies, planned for the second stage of the project, have not been released by the OECD to date.

The Council of Europe started a similar project for history teaching in Europe with a conference in Bruges, Belgium in December 1991. Perhaps a more detailed report with information about the actual situation in history teaching, covering objectives, content, timetables, didactical arrangements, results, may be published soon. One of the joint projects of CIDREE is starting this year with a provisional comparative study in history teaching. There is an interesting study by Crum (1991) of the prescribed curriculum in physical education in various European and North American countries. We all know of the valuable work of the IEA on mathematics education, science education, computer education and writing. The reports, some of which have already been mentioned, contain a lot of information about the situation in these subjects.

In the Netherlands Pieters (1990) published a similar review of the situation in environmental education and of course there is a lot of other material. But accessibility is difficult. Language problems are very important and the level of specification is very different. It is promising that in various countries more attention is being given to national assessments of educational progress. This is borne out not only by the reports on learning results, but also by those on the actual taught curriculum as investigated by questionnaires for teachers and by observation and analysis. A very interesting approach may be found in the work of the IMEN: the International Mother Tongue Education Network. See for example Haneis and Herrlitz (1991).

Systems for national assessment have been developed in the Netherlands and Sweden in the last few years. This occurred earlier in England and Wales (but the Assessment of Performance Unit APU has been broken down into more or less isolated projects). France introduced a similar system in 1990 and Spain is in the process of doing so.

It would be worthwhile if on a European scale these national systems for educational assessment cooperated in using each other's instruments and methodologies, preferably also in conjunction with IEA for the purpose of drawing up reports about the situation in certain subjects on a European scale. Then in a couple of years we could discuss the curricular situation in European countries on a more solid information base.

A European centre for comparative curricular information could also be very important for the further development of curricular renewal in Europe. Such a centre could stimulate international comparisons in a coordinated manner and it should be very important to make information from various countries accessible. Even on a national scale information about the curriculum is often inaccessible. It is important not only to carry out the documentation work (EUDISED and other systems) but also to publish real comparisons of, for example, lists of attainment targets for certain subjects, research findings about learning results and the taught curriculum, analysis of textbooks (as is done already for Germany by the Georg Eckert Institut in Braunschweig), etc.

Although I don't have a sound information base, I think it is safe to say that in no country has the curricular situation in schools not changed in the last twenty years. Perhaps the changes are not as dramatic as was hoped by curriculum specialists and educational scientists. This is probably the background of the pessimistic view taken by Goodlad, Cohen and Spillane. But nevertheless a lot of small innovations have been introduced in Dutch and European schools, not just in textbooks, but also in examinations, teaching behaviour, learning behaviour, communication between teachers and students and in the content of subjects like mathematics, science, computer literacy, foreign language teaching, the mother tongue, history, arts. In the design, elaboration and implementation of these important changes in subjects, curriculum research has played an important role in the Netherlands. This is borne out in universities linked up with projects of the SLO and in teacher training colleges, looking into the training itself, the local guidance centres and the interaction between research groups, SLO teams and implementation teams from local guidance centres and teacher training colleges. Although much of this investment did not reach the average school and the average teacher, some of it did filter down to almost all schools, certain innovations being implemented on a large scale. This is the case, for example, with new approaches in the mathematisation of everyday situations in mathematics teaching; with fieldwork and school-based experiments in science teaching, with the emphasis on communication patterns in foreign language teaching, and the structured development of creativity in arts and crafts, stimulated by the introduction of these subjects as examinable electives. One may also cite the introduction of a new subject 'political and civic education' for all students, highlighting local democratic decision-making, societal problems, ethical problems etc. Dutch schools and Dutch school curricula in secondary education have come a long way since 1970. My many contacts give me reason to believe that the same is true in other European countries. But changes are slow and piecemeal and not very well documented.

It is impossible in the framework of this paper to go into details and to document all the changes that have been investigated by educational research initiatives in hundreds of small reports, besides those that have featured in reports of the Inspectorate, examination committees and evaluation studies of SLO projects.

A specific point of interest is the infusion approach for the cross-curricular issues and topics. I have already mentioned similar approaches in, for example, the Northern Ireland Curriculum Council NICC, the Landesinstitut für Erziehung und Unterricht LEU in Baden Württemberg, the Libero Blanco in Spain, the new Gymnasium curriculum for Bavaria and the work of groups in France (INRP) and in the SLO. In the Netherlands, this type of work goes hand in hand with curriculum research. One example will illustrate the approach: consumer education was seen (ca 1980) as important in the affluent society in which also young children had enough money to develop consumer behaviour. Research investigated patterns of choice of younger and older children in spending their pocket money and decisions of adults on spending money. Logic based decision-making was compared with the results of

these investigations. Whereupon SLO specialists designed a curriculum in terms of objectives and contents, experimental situations, concepts etc. So a 'consumer education' curriculum was designed. It was also published, but not for immediate implementation, because it was very clear that the incorporation of a new subject 'consumer education' was impossible within the structure of traditional subjects. So the curriculum was broken down into suggestions for the existing curricula in biology, economics, physics, Dutch language and mathematics. Specified lists of recommendations were drawn up for textbook developers, in-service trainers and teachers using textbooks for traditional subjects. In these lists the content of the cross-curricular issue 'consumer education' was covered in order to facilitate the infusion approach in the existing curriculum. The same procedure was also followed for environmental education, the European dimension, developmental education and peace education. As some evaluation research on the use of this type of material has shown, two things are essential:
-     teachers must be able to find in a very short text the heart of the idea of the cross-curricular issue ('what is the essential element in consumer education?') and to grasp some ideas of possible activities, content and didactical arrangements;
-     teachers have to find instruments and put forward suggestions that can easily be infused in their existing curriculum and their textbooks.

There is also the central problem of horizontal and vertical coordination in the curriculum for secondary schools. This problem receives too little interest in curriculum research, curriculum development and management development. Curriculum coordination in secondary schools is not developed and although research on effective schools shows that a strong curricular and educational leadership is essential for good results, European schools are weak at this point, because management is mainly based on an administrative tradition. Therefore it is very important to publish very practical suggestions for individual teachers delivering certain subjects or teachers cooperating in a subject team in order to help them bring about this horizontal and vertical coordination in the curriculum. Teachers are encouraged to make contact with their colleagues from other departments, who, perhaps, want to make use of the same framework and possibly the same practical recommendations in dealing with a specific cross-curricular issue. This approach in curriculum development is in most cases not based on solid research about planning and coordination in schools, but on common knowledge. In the last 10 years or so the approach has been enforced by regulations on cross-curricular issues with lists of attainment targets prescribed by governments. But there still is very little research on the effect of this approach on the content of textbooks for different subjects and on the practice of teaching.

CONCLUDING REMARKS

It was impossible to sketch a panoramic view of educational research on secondary school curricula in the Netherlands, let alone on a European scale. Nevertheless some

77

lines of research and their results have been given. These are summarized in the following statements:

1. Research and theory failed to design a new structure for a comprehensive junior secondary education and to develop and implement a curriculum based on such a 'grand design'. Various evaluation reports have shown that there is not much correlation between the structure of schools and the results obtained, although in terms of cultural and social education and opportunities for young children to learn to cooperate over the barriers of socio-economic strata, a moderate system of comprehensive education seems to be the best.

2. But all the investment in the design of new systems of comprehensive education, including new curricula, perhaps based on new thinking about culture and society (Lawton, Von Hentig et al) has had its influence on the gradual curricular change in subjects: the fading out of sharp boundaries between subjects; the infusion of new content in some subjects and the package deal on a system of moderate comprehensive education that has been developed in the last five years in almost all European countries. There is a lot of research connected with small-scale curriculum development projects, that has contributed considerably to the investigation of new ideas for larger or smaller domains of the curriculum. Small-scale innovation of content and didactics has been successful in many subjects and topics. We still have problems with large-scale innovation. There is a very slow impact of this type of curriculum research and development on the everyday practice of schools and textbooks; but there is some impact nevertheless.

A summary of these important changes in the secondary school curriculum in the Netherlands, as it is delivered in reality, may be found in Van den Brink and Van Bruggen (1990) and in a paper of Van Bruggen entitled 'Survey of Trends in Curriculum Reform in the Netherlands'(1987) written for the OECD project 'The Curriculum Redefined' (see Skillbeck, 1990).

3. A serious shortcoming is the absence of an information centre on a European scale for the exchange of results of curriculum research and curriculum development in practical and detailed areas of the curriculum. Such a centre could be a great help in improving the quality of the gradual introduction of the curriculum. Such a centre could also help in making the national systems of educational assessment more efficient and provide information about what is actually going on in European schools.

4. There is a lot of research about general educational problems like motivation, learning theory, development of values and character, etc. This type of research has had very little influence on curriculum development and still less on the curriculum practice of teachers and textbook developers. A good example may be found in a number of articles about 'learning to learn' in a special issue of the journal "Schularbeiten" (Schoolwork) of the Landesinstitut für Schule und Weiterbildung

LSW (1990) in Soest, North Rhine-Westphalia, Germany. Nevertheless this influence could be increased, if curriculum designers in cooperation with researchers from these areas could find time and money in order to infuse provisional results from this general research into their very practical work on the curriculum. That this type of work is not easy is very well illustrated in the report of the OECD Conference 'Learning to think, thinking to learn' (1989) - see Maclure (1990) - where various promising approaches for the promotion of a 'thinking curriculum' were presented. These were partially based on specific courses (De Bono and others) and partially on an infusion approach into usual subjects.

5. It has appeared to be very difficult to base curricular decisions on a scientific analysis of the needs of society and of future young adults, who are now students in junior secondary schools. There are not only huge methodological problems, but it has also appeared that this type of thinking and research (future analysis; trying to formulate possible consequences etc.; compare Rassekh and Vaideanu and others) is too general and too hypothetical. In some recent curriculum development projects an attempt has been made to seek the opinions of 'society' (parents, representatives of societal groups, churches, employers, journalists ...) about proposals for new curricula with regard to both individual subjects, especially in vocational programmes, and the curriculum as a whole. This has been done in Bavaria in the last few years and in the Netherlands in 1985 in connection with the proposed curriculum for a comprehensive school for 12-16 that has not yet been introduced. In both instances the results of the investigation were poor. Most people only look for certain details or ask for the inclusion of new subjects or coverage of more curricular themes.

6. The traditional place of traditional subjects, upheld by teachers' associations, examination structures, timetables, qualification systems for teachers, etc is so strong that it would be stupid to try and destroy this valuable structure. Yet there are problems raised by associations or by a lack of co-ordination in school programmes as shown by Haft and Hopmann (1986) in their investigation of the work of curriculum committees in Western Germany. This is exactly what has been done in the package deals on the curriculum for junior secondary education in the last couple of years.

7. On both a European and a national Dutch scale, we have far too little research on the actually taught and learned curriculum; however systems of national assessment are emerging now. It would be wise to coordinate this type of research on a European scale. (Compare also with 3).

8. The new curricula, prescribed during the last few years for junior secondary education, offer a good opportunity to curriculum researchers and developers for rendering a service to their educational systems by taking up patient and thorough research in classrooms and study rooms in connection with teacher training and curriculum development. In my opinion, this research should focus on topics that are relevant to a gradual improvement of curricula (textbooks, assessment materials and

examinations included) and classroom practice in accordance with the main goals of the new curricula, in particular:

a.   Elaborating a series of longitudinally planned classroom actvities, that are meant to contribute to the attainment of general goals,such as the development of study skills, thinking behaviour, etc. within the curricula for prescribed subjects, which are based in part on psychological theories of these skills and aptitudes; on creative design of activities; and on patient formative evaluation on a small scale, together with a strong emphasis on 'infusion' in textbooks, examinations and manuals.

b.   The same type of research and development for cross-curricular themes.

c.   Research in the area of informal, incidental and parallel learning of students outside schools. What are the knowledge, skills and attitudes that students have acquired about, for example, the significance of science and technology in our modern life from watching TV etc.? What are the most important differences between various groups (sex, social and cultural class)? And how can differentiated curricular offerings profit from this learning and from the differences?

d.   In various subjects and domains of the curricula, what are the essential facts, concepts, principles, necessary for a smooth transition to tertiary education? Careful analysis of transition problems may help in improving the efficiency of longitudinal planning of curricula and may help focus the debate on 'what is essential' in, for example, the chemistry curriculum for 15-17 year-olds with a view to their acquiring 'general knowledge' (cultural literacy) together with the specialized knowledge and skills required for further study in various courses.

e.   Research on the possibility of offering a common curriculum to 11-15 year-olds for most subjects and topics, but with built-in differentiation in learning modes; variations in learning results but with enough commonality in order to continue with the whole class; using intellectual and social differences among students as positive elements for comprehensive learning and teaching (see, for example, the promising experiment of Herfs, 1991 and others for mathematics teaching with 12-15 year-olds). This differentiation is one of the most persistent problems in the junior secondary curriculum and general solutions, such as tracking, streaming and intra-class differentation based on interest, raise problems.

Of course, research on these topics is in hand. But more international exchange and cooperation could help make results more accessible to developers of curricular materials, textbooks, examinations, in-service-training courses etc. It could also help by focussing research on practical problems. (In the elaboration and implementation of the new package deals on the curriculum for 11/12 and 15/16 year-olds).

9. One of the most important problems is the vertical coordination in the curriculum between primary, junior secondary and upper secondary schools. Many problems still exist in terms of efficiency and effectiveness. The same is true of the horizontal coordination between subjects delivered by different teachers in junior secondary schools. Curriculum management at department and school level will become more

and more important during the coming years as a means of promoting the flexibility and freedom that are important elements of the package deal in the junior secondary curriculum. The development of further procedures and instruments for school managers, based on research close to the real practice in schools, will grow in importance.

10. In many curriculum projects, aiming at lists of attainment targets in various countries, an important issue has been the definition of the core of knowledge, skills, insights and attitudes in a particular subject. In the preparation of the Gymnasium curriculum in Bavaria this has been a very important feature involving the identification of the so-called 'Grundwissen' in all subjects. But the same was the case in the working groups in England, Spain and the Netherlands and is now the case in France. Further analysis and investigation and solid scientific subject-based research combined with psychological and didactical research in schools will remain very important for helping schools and governments in the selection of the core in the package deal of the curriculum.

REFERENCES

Akker, J.J.H. van den (1988): Ontwerp en implemntatie van natuuronderwijs (Curriculum-design and implementation of science in primary education). Swets and Zeitlinger, Lisse, the Netherlands.

Aurin, K. (1986): Gegliederten Schulsystem und Gesamtschule: Vergleichsuntersuchung des Landes Baden-Württemberg (3 Teile). Ministerium für Kultus und Sport, Stuttgart, Germany.

Ball, S.J. and S. Larsson (1989): The struggle for democratic education: equality and participation in Sweden. Falmer Press, New York, USA.

Bayerische Staatsministerien für Unterricht und Kultus und Wissenschaft und Kunst (1990): Lehrplan für das Bayerische Gymnasium. Munich, Germany.

Berg, R. van den, U. Hameyer, K. Stokking (eds.) (1989): Dissemination Reconsidered: The Demands of Implementation. ACCO, Louvain, Belgium.

Bouwens, F. and M. Oud-de Glas (1991): Het vreemde-talen-onderwijs in de onderbouw van het voortgezet onderwijs (Foreign Language Teaching in Junior Secondary Education). Swets and Zeitlinger, Amsterdam, the Netherlands.

Bourdieu, P. and F. Gros (1989): Principes pour une reflexion sur les contenus d'enseignement. Le Monde de l'Education, April 1989. Paris, France.

Brink, G.J. van den and J.C. van Bruggen (1990): Dutch curriculum reforms in the

1980s. In: The Curriculum Journal, Vol. I, issue 3.

Brink, G.J. van den and H. Hooghoff (eds.) (1990): Core curriculum for basic education in Western Europe: perspectives and implications. Consortium of Institutions for Development and Research in Education in Europe CIDREE. SLO, Enschede, the Netherlands.

Brink, G.J. van den (1992): Dutch proposals on national attainment targets for the 15+. SLO, Enschede, the Netherlands.

Bruggen, J.C. van (1986): Establishing a modern core curriculum: a tricky business or a political art. In: Gorter, R.J. (ed.): Views on core curriculum; contributions to an international seminar. SLO, Enschede, the Netherlands.

Bruggen, J.C. van (1987): Survey of Trends in Curriculum Reform in the Netherlands. SLO, Enschede, the Netherlands.

Bruggen, J.C. van (ed.) (1987): The problems and possibilities of the evaluation of institutes for curriculum development. Special issue: Vol. 13 nr. 3. Studies in Educational Evaluation. Pergamon Press, Oxford, UK.

Bruggen, J.C. van (1989): The creation of curriculum knowledge for, by and with autonomous users: the case of curriculum development. In: Berg, R. van den, U. Hameyer, K. Stokking (1989).

Bruggen, J.C. van (1992): Hoe gebruikt het onderwijs de SLO?: samenvatting en analyse van onderzoeken naar de doorwerking van SLO-projecten. (How is education using the SLO?: summary and analysis of investigations into the impact of SLO projects.) SLO, Enschede, the Netherlands.

Cohen, D.K. and J.P. Spillane (1992): Policy and Practice: the Relations between Governance and Instruction. In: G. Grant (ed.) (1992): Review of Research in Education 18. AERA, Washington, USA.

Coulby, D. and L. Bash (ed.) (1991): Contradiction and Conflict: the 1988 Education Act in action. Cassell, London, England.

Crum, B. (1991): Physical education as part of the core curriculum in secondary education: a restricted, comparative international survey. SLO, Enschede, the Netherlands.

Fend, H. (1982): Gesamtschule und Vergleich. Bilanz der Ergebnisse des Gesamtschulversuchs. Beltz, Weinheim, Germany.

Flude, M. and M. Hammer (ed.) (1990): The Education Reform Act 1988: its origins and implications. Falmer Press, London, England.

Gehrke, N.J., M.S. Knapp, K.A. Sirotnik (1992): In Search of the School Curriculum (pages 51-110). In: G. Grant (ed.) (1992): Review of Research in Education 18. AERA, Washington, USA.

Glatthorn, A.A. (1987): Curriculum Renewal. Association for Supervision and Curriculum Development ASCD. Alexandria, VA. USA.

Goodlad, J.I. (1984): A Place called School: Prospects for the Future. McGraw-Hill, New York, USA.

Grant, G. (ed.) (1992): Review of Research in Education 18. AERA, Washington, USA.

Haft, H., S. Hopmann a.o. (1986):; Lehrplanarbeit in Kommissionen. Institut für die Pädagogik der Naturwissenschaften. Kiel, Germany.

Hameyer, U. (1987): AKTIF-Erfahrungsberichte und Studien (AKTIF-Reports from experience, studies). Institut für die Pädagogik der Naturwissenschaften IPN, Kiel, Germany.

Hameyer, U. (1992): Stand der Curriculumforschung; Bilanz eines Jahrzents (Curriculum Research, State of the Art). In: Unterrichtswissenschaften nr.3. Juventa Verlag, Munich, Germany.

Haneis, E. and W. Herrlitz (eds.) (1991): Comparative studies in European standard language teaching; methodological problems of an interpretative approach. International Mother Tongue Education Network IMEN. SLO, Enschede, the Netherlands.

Hargreaves, D.H. (1982): The Challenge for the comprehensive school. Culture, curriculum and community. Routledge and Kegan Paul. London, England.

Hentig, H. von (1971): Allgemeine Lernziele der Gesamtschule. In: Lernziele der Gesamtschule, Deutsche Bildungsrat, Klett, Stuttgart, Germany.

Herfs, P.G.P. a.o. (1991): Leren door samenwerken: adaptief groepsonderwijs als een curriculuminnovatie in het voortgezet onderwijs (Cooperative learning and adaptive instruction in mixed-ability groups in secondary education). Swets and Zeitlinger, Amsterdam, the Netherlands.

Hirsch, E.D. jr. (1987): Cultural Literacy; what every American needs to know. Houghton Mifflin Company, Boston, USA.

Kuiper, W.A.J.M. and A. Alting (1990): Biologie, natuurkunde, scheikunde en kennis der natuur: de feitelijke lespraktijk in beeld: eindrapport (Biology, physics, chemistry, knowledge of nature: the taught curriculum investigated: final report). OCTO, University of Twente, Enschede, the Netherlands.

Kuiper, W.A.J.M., E.M. Kuhne, J.J.H. van den Akker (1989): De implementatie van een maatschappijleer-curriculum (The implementation of a curriculum for civics). University of Twente, OCTO, Enschede, the Netherlands.

Lawton. D. (1975): Class, Culture and the Curriculum. Routledge and Kegan Paul, London, UK.

Lawton. D. (1982): The End of the 'Secret Garden'? A study in the Politics of the Curriculum. University of London Institute of Education.

Landesinstitut für Schule und Weiterbildung (1990)? Lernen lernen (Learning to learn). Special issue of 'Schularbeiten', Heft 3, November 1990. LSW, Soest, Germany.

Lieberman, A. (ed.) (1986): Rethinking School Improvement; Research, Craft and Concept. Teachers College Press. New York, USA.

Lewy, A. (ed.) (1991): The International Encyclopedia of Curriculum. Pergamon Press, Oxford, UK.

Maclure, S. (1990): Learning to think, thinking to learn. Pergamon Press, Oxford, UK.

Marklund, S. (1980-1984): Skolsverige 1950-1975 (4 volumes) (Educational Reforms in Sweden 1950-1975). Stockholm' Liber-Utbildningsforlaget, Stockholm, Sweden.

Ministère de l'Education Nationale (1991a): Propositions du Conseil National des Programmes pour l'Évolution du Collége. Paris, France.

Ministère de l'Education Nationale de la Jeunesse et des Sports (1991b): Quel Lycée pour domain? Propositions du Conseil National sur l'évolution du lycée. Paris, France.

Ministerio de Educacion y Ciencia (1990a): Diseño Curricular Base; Educacion Secundaria Obligatoria I. Madrid, Spain.
Ministerio de Educacion y Ciencia (1990b): The White Paper: For the Reform of the Educational System. Madrid, Spain.

Ministerium für Kultus und Sport, Baden Württemberg (1984a): Bildungsplan für das Gymnasium Band I und Band II. Stuttgart, Germany.

Ministerium für Kultus und Sport Baden Württemberg (1984b): Bildungsplan fur die Realschule. Stuttgart, Germany.

Neave, G. (1992): The Teaching Nation (forthcoming).

Northern Ireland Council for Educational Development (1987): The Transition to Adult and Working Life Project (TRAWL); end of project report. Belfast, Northern Ireland.

Pieters, M. (ed.) (1990): Teaching for sustainable development: report on a workshop at Veldhoven-Netherlands 23-25 April 1990, including starting points for environmental education: a design for a core curriculum in environmental education. SLO, Enschede, the Netherlands.

Postlethwaite, T. and D.E. Wiley (1991): The IEA Study of science II: science achievement in twenty-three countries. Pergamon Press, Oxford, England.

Rassekh, S. and G. Vaideanu (1987): The contents of Education; a worldwide view of their development from the present to the year 2000. Unesco, Paris, France.

Robinsohn, S.B. (1967): Bildungsreform als Revision des Curriculum. Neuweid, Germany.

Rosier, M.J. and J.P. Keeves (1991): The IEA Study of science I: science education and curricula in twenty-three countries. Pergamon Press, Oxford, England.

Schirp, H. (1987): Hauptschule und Lehrplanarbeit; Ansätze und Aufregungen zur inneren Schulreform. (Hauptschule and curriculum work; ideas for school development). Landesinstitut für Schule und Weiterbildung, Soest, Germany.

Skillbeck, M. (1990): Curriculum Reform. An overview of trends. OECD, Paris, France.

Staatsinstitut fur Schulpädagogik und Bildungsforschung ISB (1990): Materialien zum Lehrplan fur das Bayerische Gymnasium (ISB intern). ISB, Munich, Germany.

Standaert, R. (1991): De vlag in de top; over de rationaliteit van het secundair onderwijsbeleid (The banner at the top; about rationality in educational policy for junior secondary education). ACCO, Louvain, Belgium.

Stoks, G. (1989): A survey of curricula for modern languages at upper secondary school level in sixteen European countries: with special reference to the formulation of objectives and modular schemes. SLO, Enschede, the Netherlands.

Streumer, J.N., B.G. Doornekamp and L.W.F. Bonekamp (1987): Techniek in het voortgezet onderwijs. Onderzoek naar de prestaties leerlingen in het LBO, AVO en VWO (Technology in secondary education. Assessment of pupils' results in lower vocational schools and three types of schools for general education). SVO, The Hague, the Netherlands.

Weering, B. van and Tj. Plomp (1991): Information Literacy in secondary education in the Netherlands: the new curriculum. In: Computer education. Vol. 16 nr. 1.

Wetenschappelijke Raad voor het Regeringsbeleid (Dutch Scientific Council for Governmental Policy, 1986): Basisvorming (Basic Education). The Hague, the Netherlands.

Wielemans, W. and I. Vermeerbergen (1990): Basisvorming in het voortgezet onderwijs (Basic education in secondary education). Garant, Louvain, Belgium.

Wierstra, R.F.A. (1990): Natuurkunde-onderwijs tussen leefwereld en vakstructuur (Teaching physics between the daily life world of pupils and the world of theoretical concepts); een evaluatie-onderzoek aan de MAO- en HAVO-curricula van het Project Leerpakket Ontwikkeling Natuurkunde PLON (evaluation of curricula of the PLON project). CDB Press, University of Utrecht, the Netherlands.

WILL THE TWAIN MEET?
THE RELATIONSHIP BETWEEN THE WORLD
OF WORK AND THE WORLD OF SCHOOLS

Liv Mjelde
National College of Education for
Vocational and Technical Teachers - Olso, Norway

INTRODUCTION

The relationship between the world of work and the world of school has come into focus in a new way in this time of economic transition and industrial change. Vocational education, whether it is integrated into the secondary school system, or is obtained by means of an apprenticeship in a kindergarten or in the plastics industry, is at the centre of discussions in the educational reforms currently taking place in Scandinavia.[1]

The relationship between intellectual and manual labour, between the work of the hand and the work of the mind, between vocational learning traditions and academic learning traditions, has been the underlying concern in all school reforms in Norway since the Second World War. A policy of increasing and developing public schooling to attenuate class contradictions through standard universal education has been characteristic of school systems throughout social democratic Scandinavia in the post-war era. A major concern of this policy has been the capacity of schooling to promote upward social mobility among the so-called disadvantaged. Another concern has been the unification of different types of school under common legislation.[2]

Norway's school system is public; few private schools exist. It is a three-level system consisting of:
Level 1.   Primary school, consisting of a nine-year comprehensive programme (ages 7 to 16);
Level 2.   Upper secondary education (ages 16 to 19), composed of the gymnasia (the university stream) and vocational schools;
Level 3.   Higher education, comprising colleges and universities.

In 1959, Norway integrated the old secondary modern school and the grammar school into nine years of comprehensive primary school. The School Commission of 1965 was created with the mandate to carry out a systematic study of the upper secondary school system. The commission proposed that general and vocational studies be administered by the same act and organized under the same roof. All learning should take place within a unitary school, where vocational and academic studies would be integrated, in terms of both content and physical organization. Apprenticeship would be abolished. In 1974, the traditional gymnasia and the vocational schools came under one comprehensive law (The Upper Secondary School Act). A fundamental aim of the new law was to give equal status to practical and theoretical education (Ministry of Church and Education 1982:19). Another aim was to give all young people equal opportunities, regardless of class background, geographical location or gender.

The principal aim of more recent reforms of the secondary school curricula is to further integrate general and vocational education. The White Paper (No. 33 1991-1992) on educational reforms states that upper secondary education is moulded by two major traditions, stemming from the middle ages. Vocational education and apprenticeship originate in the guild system, where learning for one's working life took place in the midst of work. The learning tradition of the gymnasium has its roots in the old Latin school, with its more conceptual and theoretical traditions, based outside the world of work. The aim of the present school reforms is to make the twain meet. How this is to come about remains the central question. What kind of teaching methods will be promoted to achieve this aim: the pedagogy that is predominant in the workshop tradition of vocational education, or the pedagogy of traditional formal schools? How can learning be meaningful to all students? The aim of this paper is to explore these questions from the point of view of vocational students and apprentices, how they were affected by the school reforms of the 1970s, seen in the light of the intentions of the present school reforms.

VOCATIONAL AND GENERAL EDUCATION IN THE SCHOOL REFORMS OF THE PAST TWENTY YEARS

The integration of vocational and general education under a common law during the past twenty years has resulted in both vocational training and academic programmes perpetuating their respective traditions. In this time period, there has been a great expansion of upper secondary education, with numbers of students steadily increasing since 1976.[3] High unemployment rates among young people result in growing numbers turning to higher education, and one way the government has tried to tackle the situation is to expand the educational sector.

This was also a period of increasing differences and rivalry in educational politics between school-based vocational training and apprenticeship. While the schools were subject to political control through public administration, apprenticeship training was controlled jointly by management and labour unions. The first Apprenticeship Training Act in Norway was passed in 1950, and was revised in 1980 as the Vocational Training Act. The intention of the school reforms of the 1970s to place all vocational training within the school system was not realized in practice. Apprenticeship continued not only to live its own life during this period, but also to grow stronger, despite political intentions to phase it out.

One hundred and ninety trades are recognised under the Vocational Training Act today, and a further fifty trades are in the process of preparation. Trades that vanished in the 1960s, such as shoemaking, are reappearing; new trades are being created with burgeoning new industries like plastics. The service sector is also turning to apprenticeship training programmes for their staff.

During the 1960s and 1970s the government seemed to play the predominant part in shaping educational policies in Norway. One of the main conflicts in the 1970s between the government and the two main organizations in industry, labour unions and management, was over the relationship between the world of work and the world of schooling. In the 1980s and 1990s, there has been a tendency on the part of the organizations to play a more active role in educational reforms.

Some of the characteristics of Norwegian vocational education today are: first, the students in the vocational fields meet fierce competition for the few places that exist in the advanced courses in upper secondary school. Eleven per cent of Norwegian students in basic vocational courses within crafts and industry have a possibility of entering a second year, and a mere three per cent go on to the third year. Secondly, a number of students "drift" horizontally in the system; they take one foundation course after another, which does not lead to a professional status or matriculation rights. Thirdly, apprenticeship places follow the ebb and flow of the labour market. The competition for apprenticeships is fierce.

These problems were the focus of an important document which drew up policy directions for the 1990s (Norwegian Official Reports No. 4 1991). The policy report emphasized that basic human needs can only be satisfied through industrial production. The two main questions are: which areas of production will develop in Norway and what kind of skills will the labour market need in the future? Throughout the industrialized world, recent decades have brought a tremendous upsurge in the political as well as the economic significance of skill. This stress on skill is central to the policy report, particularly in relation to the links between vocational and general education.

The report proposes both guaranteeing the right of all young people to achieve matriculation and/or vocational qualifications, and establishing a more flexible structure in education to provide more efficient utilization and coordination of resources.

The production crisis in general and the rising unemployment among youth form a backdrop to these reforms. The proposals are also a consequence of the difficulties that arose in meeting The Young Peoples' Guarantee of 1988/89, a policy document which sees the provision of young people with either an offer of work or education as a human right.[4] Another aspect of these problems is the fact that research has highlighted a learning crisis in the educational system and has exposed the very real remoteness of the schools from the world of work. Pedagogical research has shown that the problems of meaningfulness and motivation in learning are core questions.

MEANING AND MOTIVATION IN EDUCATION

In order to gain an understanding of what creates good learning situations for students, and to understand how teaching problems emerge and change for the teachers, it is imperative not simply to study the content of education and how it is distributed and reformed, but also to study how both students and teachers perceive and evaluate the content of education and what aspects of education they select as meaningful and how they select them (Nilsson 1986).

In "The Apprenticeship Project 1982-84" (Mjelde 1990), 89 per cent of the 1617 apprentices in the survey preferred to learn at the workplace instead of at school. "Tired of school", "to do something real", and "to do practical work instead of fiddling around with theory" were typical responses students made to open questions. Research in Scandinavia has established that the school creates difficulties for 20 per cent of the students in the 7th-9th grades in primary school (Axelsson 1989:79, Mjelde, 1984, 1987). Beginning in the 1960s, many so-called disadvantaged youth have been offered places in the vocational school system as part of rehabilitation programmes. One of the difficulties shared by these young people was their frustration with the nine-year comprehensive school. The results of a study carried out on three cohorts of integrated students in the vocational school system in the 1970s showed that 70 per cent completed their course work and were either in further education or in a work situation by the time the evaluation was made (Mjelde and Hammer 1979). It was the workshop learning in the vocational school which gave them a new opportunity. An apprentice in the 1982-84 project said:

"School is not my thing in life; going to school in today's society is exhausting."

School drop-out, a problem all over the western world in the post-war era, is related to the problem of motivation. Many students in school today have lost the inclination to learn at the same time as they have lost confidence in their ability to learn. This is a crisis in motivation. Regi Enerstvedt (1987) points out that the purpose of any learning process is preparation for some subsequent activity. When school and work are as far apart as they are in the Norwegian elementary school, the question of meaning and motivation becomes a basic one for students. The academic educational model, which is prevalent in the nine-year comprehensive school, is based mainly on a value system which places theory before practice and underestimates a practical orientation.

The advantage of vocational education has been the traditional pedagogy based on learning from practice to theory and the closeness to work life. The distinctive stamp of vocational training is that the work of the mind is formed from the work of the hand, and that practice is superior to theory. Historically, the model for vocational training in Norway is found within fishing and farming occupations as well as trade and industry.

Training takes place in the midst of production; knowledge passes from parent to child, from craftsman to apprentice. Understanding comes through action and personal experience, and theory is learned in close relationship to practical skills. The vocational school system is based on this model. The students do something real: they bake bread, make sausages, build a house, cut hair, repair radios. Learning takes place amidst noisy machinery or hot stoves; students work together to perform tasks, while the teacher goes from group to group giving instructions (Berner 1989, Mjelde 1990).

Workshop teachers in the vocational school system have different backgrounds from their counterparts in general education (Hostmark-Tarrou, 1988).[5] Vocational teachers have gained their competence as manual workers in trade and industry. Their school background is mainly primary school, after which they have been through vocational training and apprenticeships. Their performance as teachers is marked by these experiences.

Who are today's vocational school students? Their ages range from 16 to 25, but most students are between 16 and 19, 60 per cent come from a working-class background of unskilled/skilled labour, 30 per cent from a petit-bourgeois background, and 10 per cent from the upper middle-class (Fleischer 1977, Mjelde 1990).

A traditional sexual division in different trades continues to be a trait of vocational education, and reflects the sex segregation of the manual labour market and society as a whole. Most girls enrol in courses which provide basic short-term education in clerical work, home economics, arts and crafts, sewing, hairdressing, nursing and auxiliary education. In terms of opportunities in the labour market, this means that girls qualify for clerical and sales work and, to some degree, for work on the lowest level of the health and education sectors (e.g. child care).[6]

This period of transition and change, however, has also brought new perspectives on the gender question in Norway, and it is having effect on educational policies. Parliament has passed a law on equality between the sexes and a Royal Commission has treated the question of the male role. An important debate is how to break down gender divisions in education and prepare men for family and caring responsibilities.

LEARNING AND VOCATIONAL TRAINING

Recent approaches to learning in cognitive science have placed central importance on interaction in the learning process. Learning is seen, not as something which happens inside a student's head when she/he listens to a teacher or studies a book, but as fundamentally interactive. Learning is the internalization of schemata which incorporate cognition, perception and action. Schemata are made meaningful by jointly carrying out activities by an "expert" in such a way that the learner gradually masters successively more difficult parts of the task. Gradually, she/he is able to take

Figure 1

Vocational Education		
Vocational Techniques		
Practical Skills	Vocational Theory	General Knowledge
Training in the use of tools in practical application	Knowledge of principle in tool use (effectively, applied physics and mathematics;) draughting; mechanical drawing and the composition of materials, (effectively, chemistry and mathematics).	Norwegian/Social science/Mathematics Physics Chemistry

over more complex stages until she/he no longer needs the expert to assist. Ordinary learning is based on joint activity leading to gradual mastery by the novice (Goody, 1989:253).

The fit with vocational learning is immediate. The central aspect of vocational training is learning by doing, gaining professional skills while interacting with materials, with teachers and with fellow worker-learners. One learns through the basic work operation, through manual practice. Here, the dynamic between theory and practice moves from the practical to the theoretical; hand and mind are in a dialectical relationship, and interaction with the teacher and other students is at the foundation of the learning process. It is apparent that the vocational school system carries within it distinct cultural values. When a teacher and twelve students spend twenty hours together in the workshop, teacher/student and student/student interaction becomes a different form of communication and learning from what is normally found in twenty hours of classroom work. In the practical work situation, emphasis is placed on reciprocal interdependence. Traditional school values, such as the notion that cooperation is cheating, are not part of the value system of vocational training. Before the Upper Secondary School Act in 1974, no grades were given in vocational subjects, only a pass/fail system was used.

However, contradictions between the work of the hand and the mind also exist within the various vocational education fields. The three components of vocational education are: practical skills, vocational theory, and general knowledge (Nilsson 1981: 37).

Vocational theory, taught both in the workshop and in the classrooms, has varied over the decades, depending on the subject-matter and the historical period. Draughting is an essential part of vocational theory. Vocational theory is linked to the content and development of particular subjects. Its purpose is to impart knowledge about tools and materials used in each field. General theory, however, has been separated from practical skills and is not linked to any special field or subject. This dichotomy has led to the theoretical impoverishment of vocational pedagogy for over a century (Jackson 1992). In the words of an electrician in the Apprenticeship Project 1982 - 1984 "The school cannot give you realistic training - neither for a craftsman, nor for an academic". (Mjelde, 1990:43)

The relation between the abstract world of academic education, unaffected by practical reality and the concrete world of vocational education, with its root in practical work, (but not linked consciously to the laws of science which underlie the practical), is a classical theme, both in the sociology of knowledge and in educational policy. While the question of further integration between vocational and general studies has now emerged as a political issue, it is not the result of any theoretical clarification between these two traditions in education. It is rather an acknowledgement of the fact that developments in the sphere of technology are radically changing occupational divisions, as well as the contents and organization of work. This means more pressure to integrate the different elements in educational careers.

The main model for vocational education under the new reforms in Norway is two years spent in school, followed by a two-year apprenticeship in working life. The proposed agreement, based on the "2+2 model", proposes a cooperative system in which schools are given greater responsibility for completed training if it is shown to be impossible to find an apprenticeship place in industry.

A further topic of debate is the relationship between breadth and specialisation. The concepts in the general educational policy debate are used in relation to the sequencing of training. When and for how long is breadth/specialization to be given? How broad are the trade skills to be (trade generalist or specialist), and what should the relations between vocational/general topics be? The White Paper proposes a few broad entrances to upper secondary education and specialised exits, preferably to a specialized apprenticeship. A controversial question is whether graduates of vocational schooling will have college and university matriculation rights.

The White Paper does not, however, touch the difficult pedagogical questions involved and does not grasp the complexities of vocational education in full.

CONCLUSION

The questions I have presented here are being discussed all over the western industrialized world at this time of transition and change. There is also a search for alternative pedagogies (Grubb, Kalman, Castellano, 1991:4). Vocational training, with its distinct pedagogical model, offers other ways of thinking about learning. On the other hand, the transition needed will also open possibilities for reversing the impoverishment of vocational education.

Vocational training, however, contains a variety of different educational roads, characterized as much by their similarity as by their diversity. Similarity is found in relations to the labour market, and the only constant is continuous change. Diversity in the 180 trades, now classified under Norway's Apprentices Act, stems from different traditions within the craft system. Some date back to the guild system, while others have arisen with industrialization. Today, new trades are coming into being with further industrialization and the development of the service industries. Diversity and complexity permeate the crafts and industries, social service fields, food industries etc. Every field has its distinctive stamp, even if the pedagogical model of learning from praxis to theory is the same for all fields.[7]

Recession and increasing unemployment are forcing changes. The period of post-war development shows that it is in such spells of recession in Norway that the education system has its strongest period of growth and reform. Industry is restructuring and education becomes an important tool in the government's plan of action (The Norwegian Ministry of Education, Research and Church Affairs, 1992:50). The recent upsurge of interest among both educators and politicians in the dichotomies in educational thinking, the traditional separation between the work of the hand and the work of the mind, between praxis and theory, between vocational and general education, hold promise for transcending these dichotomies in the education of the future.

NOTES

1. This paper draws on current empirical research being undertaken by the author in the field of vocational education in Norway, on how vocational students and apprentices experience their learning in everyday life, at school or/and in the work place. I have also drawn upon my research into technological change, changes in the labour processes and in the relationship between men and women, skilled and unskilled labour in the printing industry. I am also familiar with development and research within this field in other Scandinavian countries, especially Denmark and Sweden.
2. Laws and policies governing the educational system in Norway are created by parliamentary decisions.
3. The expansion in the Norwegian school system has been enormous during the past 30 years as elsewhere. In 1956, 16 per cent of the age cohort of 14-year-olds went on to further education, eight per cent went to vocational schools and apprenticeships and eight per cent to the gymnasiums. Today, 95 per cent start further education after nine years of compulsory school, after age 16.
4. The right to work is guaranteed in the Norwegian constitution.
5. The teachers in vocational education and general education have different working hours and wage conditions. They are often members of different unions.
6. We have seen some major changes in this pattern during the past 20 years. Women are entering the skilled manual labour market. In the printing industry, the majority

of apprentices are women (Mjelde 1992).
7. I have described these learning traditions as "ideal types" of what we perceive as the academic versus the practical world. These contradictions and the pedagogical model of learning from practice to theory are also found in medicine, dentistry, nursing and architecture; all of which are craft-oriented professions (see: Becker, Howard "Schools are a Lousy Place to Learn in.")

REFERENCES

Axelsson, Rune (1982) : *Gymnasieskolans studievager,* Pedagogisk forskning, nr. 37 Uppsala

Berner, Boel (1989) : *Kunskapens vager. Teknik och larande i skola och arbetsliv.,* Arkiv forlag. Lund

Enersvedt, Regi (1987) : *Virksomhet og mening. utviklingen av laeremotivasjonn hos norske skolebarn,* Institute of Sociology, Univ. of Oslo, Oslo

Fleischer, Nina (1977) : *Ungdom og yrkesvalg,* Institute of Pedagogy University of Oslo, Oslo

Goody, Esther (1989) : "Learning, Apprenticeship and the Division of Labor" in Coy, Michael W (eds) Apprenticeship *From Theory to Method and Back Again,* State University of New York Press, Albany, N.Y.

Grubb, W. Norton, Kalman, Judy Castellano, Marisa (1991) : *Readin, Writin, and rithmetic one more time: The role of remediation in vocational education and job training programmes.* The National Centre for Research in Vocational Education, The University of California, Berkeley.

Jackson, Nancy (1992) : *Rethinking Vocational Knowledge,* Faculty of Education, McGill, Montreal

Ministry of Church and Education (1982) : *Vocational Training in Norway, Oslo*

Ministry of Education, Research and Church Affairs, (1991); *Veien viders til studie og yrkeskompetanse for alle,* Norwegian Official Report No 4 Oslo

Ministry of Education Research and Church Affairs, 1991/92: *Kunnskap og kyndighet. Om visse sider ved videregaende opplaering.,* White Paper Nr. 33 Oslo

Ministry of Education Research and Church Affairs (1992) : *The Changing Role of Vocational and Technical Education and Training,* Report OECD Oslo

Mjelde, Liv and Hammer, Torild (1979) : *Behanlingsklientell i yrkesskolen* Helse og sosialforskning Socialdept, Oslo

Mjelde, Liv (1984) : *Skolen som problemskapende faktor for dagens arbeiderungdom.* in Haaland, Thomas (eds) Ungdom og arbeidsloshet, Pax, Oslo

Mjelde, Liv (1987) : *From Hand to Mind*, Livingstone, David (eds) *Critical Pedagogy and Cultural Power*, Bergin and Garvey Publishers, Mass

Mjelde, Liv (1990) : *Labour and Learning: The Apprenticeship programme in Norway.* Interchange Vol 21, No. 4 OISE, Toronto

Mjelde, Liv (1992) : *Kjonn, arbeidsdeling of forandring i den grafiske bedriften* in Mjelde, Liv, & Tarrou, Anne-Lise Hostmark (eds): *Arbeidsdeling i en brytningstid* Ad Notam Gyldendal, Oslo

Nilsson, Lennart (1981) : *Yrkesutbildning i nutidshistorisk perspektiv*, Institute of Pedagogy, University of Goteborg, Goteborg

Nilsson, Lennart (1986) : *Fackdidaktik ur yrkespedagogiskt perspektiv* Marton, Ference (eds) *Fackdidaktik Akademieforlaget*, Lund

Tarrou, Anne-Lise Hostmark (1988) : *Challenges in the Education of Teachers of Vocational Subjects and Strategies for Confronting these Challenges*, European Journal of Teacher Education Vol 11, Nos 2/3.

Ministry of Church and Education 1982: *Vocational Training in Norway* Oslo

Ministry of Education, Research and Church Affairs, 1991: *Veien videre til studie og yrkeskompetanse for alle* Norwegian Official Report No. 4 Oslo

Ministry of Education, Research and Church Affairs, 1991/1992: *Kunnskap og kyndighet. Om visse sider ved videregaende opplaering.* White paper nr. 33 Oslo

Ministry of Education, Research and Church Affairs 1992: *The Changing Role of Vocational and Technical Education and Training.* Report OECD Oslo

REPORTS OF THE WORKING GROUPS

REPORT OF GROUP 1

Chairperson: **Mary Darmanin (Malta)**
Rapporteur: **Frank Gatt (Malta)**

Advice to Ministers:
Remind ministers that research is:
- often long term and ongoing and may not produce quick results;
- sometimes critical of current practices;
- sometimes difficult to read by virtue of its methodological and theoretical sophistication;
- often not directly amenable to policy formulation.

Ministers need:
- to understand the long-term goals of the research community;
- to use "bad news" or critique in a positive way to indicate that they are up to date and ready to make informed changes to improve systems;
- to indicate to the research community areas of study directly relevant to policy formulation;
- to commission said researches and projects;
- to ensure that their advisers are competent to adjudicate between different results;
- to establish liaison networks with the research community;
- to personally read abstracts of important research in order to have a first-hand feel for the work that is done and the arguments put forward.

Researchers need:
- to make clearer cases for the status of seemingly esoteric research;
- to write abstracts that summarize the research and indicate its applicability to policy;
- to establish links with the media including newspapers, radio and television, in order to disseminate their findings to a broad sector of the population;
- to maintain links with schools and teachers through teacher-training institutions and other ways;
- to keep up to date with popular political and social aspirations and aspirations in the educational field;
- to develop awareness of the relevance of their work to broad national and international objectives and practices.

Linking with schools:
Two main strategies can be useful:
First strategy: School and teacher-led research
Schools can:
- indicate specific areas of concern;

- provide easy access to researchers for field work;
- disseminate findings in the school and, if the project is generalizable enough, to the teaching and research community.

Some of this work should:
- be carried out in the form of action-research, either in collaboration with teachers or by teachers themselves;
- link up with work in other schools in the same school system;
- disseminate findings.

Second strategy: More academic-led research

This can:
- be oriented to current education issues in the broader sense;
- develop thinking about existing knowledge in the field which can later be applied to specific contexts;
- indicate applicability and relevance;
- be disseminated in various forms other than the more technical.

Selection of curriculum content
Criteria for selection should include:
- consideration of pupils' needs including stage, age, ability, class, sex, race and other special needs.
- consideration of the differing epistemological practices of each subject discipline.
- latitude within subjects for the inclusion of cross-curricular themes.
- the pedagogical practice relevant to those pupils and that discipline, and to teacher expertise.
- consideration of economic and social limits and aspirations.
- national system maintenance.

We find that:
- a common core is desirable.
- autonomous school-based curriculum development is also important.
- overload is a serious problem which needs to be addressed.
- interdisciplinary work is of questionable value because
  a. Often it consists of no more than subject specialists choosing common themes with little consideration of the relevance to pupils.
  b. Subject specialists need broader knowledge but cannot be expected to have the background that some interdisciplinary work requires.
  Therefore, interdisciplinary work should be used judiciously:

  - cross curricularity is important to broaden the curricular field but should not contribute to curriculum overload.
    New topics of interest such as health education, peace education and the European dimension should be integrated in this way. Thus all teachers and

pupils will develop a broader understanding of their subjects and relate them to increasingly wider issues.

Promotion of moral values:
We find that moral values are best promoted in and through their practice in social relations in the school, between the school and the community, and between Government and the community of citizens.

Individualization of learning:
We define individualization of learning as work that is geared to develop the skills, interests, abilities and learning needs of individual pupils.

We find:
- that more research is needed to understand how pupils learn;
- constructivist theories are useful in this respect.

The relation of curriculum to technical education:

Most systems have a general core which is taught in technical schools but since this is usually taught away from the Workshop it appears to:
- have little relevance to pupils.
- create problems of motivation.
- create disadvantages for these pupils.

Further investigation of ways to integrate curriculum areas for these pupils is needed.

Promoting autonomy for local schools:

- we acknowledge the importance of autonomy in curriculum development;
- we find that locally produced curricula are sometimes more relevant to schools;
- we encourage the development and maintenance of local cultures and knowledge;
- we encourage the development of the competency of teachers and heads;
- finally, we recognize the need to balance the desire and need for autonomy and diversity in curricula with the need for common cores so that equality of outcome can be promoted.

Language:

- we endorse a philosophy which encourages the study of one's own mother tongue as a **fundamental human right**, and the study of more than one language.
- we recognize the importance of dominant world languages but recommend that attention be given to the study of minority languages to promote the understanding and knowledge of diverse cultures and communities and to ensure the continued existence of those minority cultures and languages.
- **we suggest that** incentives should be developed to attract pupils and teachers into

the study of minority cultures and languages.

Education and the world of work:

We find that
- general competencies must be improved in vocational schools.
- work experience should be examined as a way of motivating pupils and of making school more relevant to them.
- research on how pupils learn should focus particularly on learning through practical application in work.

REPORT OF GROUP 2

Chairperson: Frank Ventura (Malta)
Rapporteur: Herbert Puchta (Austria)

1. Influence of research/research institutions on educational practice

1.1. Developmental research: a lot of influence via teacher training, teacher centres, teaching materials, etc.

- direct influence on classroom work through inservice training if there are "opinion makers" who are accepted by teachers and who translate into action patterns the findings of serious research.
- this seems to work best in smaller countries/systems.
- this way of multiplication needs to be sustained over a longer period (several years).
- financial and time constraints.
- school development under serious financial constraints is still possible, however greater risks have to be taken by politicians (less control).

1.2. Decision-related research: influence of research difficult to be traced here.

Advice to be given to researchers/research institutions:
- researchers have to "sell" their work better;
- have a responsibility to "go public" - parent groups;
- should do research into policy making and implementation;
- newspaper article - TV;
- should refrain from ideological positions, more "if ... then ..." reasoning needed.

Advice to be given to Ministries/Departments of Education:
- establish "strategies units" within the framework of the Ministries that have no administrative responsibility, but act as mediators between research and decision-makers, make scenarios, have conferences with developmental researchers,etc.,function: think tank.
- should make it possible that investments in educational research are sustained over the years (funding of long-term educational research projects).
- should allow for exceptions and experiments so that new ideas can be piloted (former Communist States).

2. Inter-disciplinarity

Neglected by research, research needed into the following problem areas:

2.1. If curriculum consists of subjects and cross-curricular themes, overloading of the curriculum results.

2.2. How can cross-curricular themes be woven into one coherent curricular system (coordination in curriculum and practice)?

2.3. How can curriculum management on the level of the schools avoid the same content being covered incoherently by various subjects.

2.4. Which content should be covered by which subjects? e.g. environmental issues in biology (or better philosophy, religious instruction, law, etc.)

2.5. What should be abandoned to avoid overloading?

2.6. How can interdisciplinarity be assessed/tested?

important influence of testing on teaching -

Here a national curriculum seems necessary to negotiate *what* should be taught, *how* and in *which* subjects?

And *when* are the topics to be taught? (time question)

Suggested model: a coherent and balanced teaching of global topics that aims at the avoidance of overloading the curriculum calls for:

- the logical analysis of content *and* process (sets priorities);
- a school-based coordination of what is to be taught and needs to be considered on various levels:
    - on the level of national curricula;
    - in-inservice training;
    - on the level of textbooks and other teaching materials.

3. The European dimension

- Fact finding research needed:
- What is the actual knowledge of children of one nation about other peoples in Europe?
- *Then*: what can be done to enhance the children's concepts?
- Balance of teaching of both, *knowledge* and *cultural concepts*, is important so that children can grasp the notion of diversity within Europe.
- Examples of influences, patterns of diversity need to be given as case studies - mosaic of identities/cultures.
- Balance of local/national/international identity is important; from local to national to European to global issues.
- Series of good publications for teachers about European issues in six or seven languages needed; is of vital importance, especially for small countries (information network).
- A look into research on direct experience of children with children of other nations is important.
- What creates - insights/tolerance, etc.? xenophobia?

4. School autonomy

4.1. It is important to stimulate school autonomy - it is a precondition to school development.

4.2. Autonomy is not a goal, but a means to achieve certain goals.

4.3. There are risks that must be made explicit.

4.4. It is necessary to have investment of *money* and *ideas*.

4.5. Important question of *standards*.

Standards need to be negotiated internationally.

Procedure: - determine standards, measure students' performance against these standards

(e.g. "European language portfolio" - Council of Europe)

4.6. What can be learnt from countries that have a long-standing tradition of school autonomy? e.g. The Netherlands:

- national document that gives a system of obligatory rules for school autonomy.
- schools have to work up their own curricula documents.
- these need to be shown to committee (of representatives of different interest groups) that gives advice to the schools.
- furthermore, choices that schools want to make (e.g. choice of subjects) have to be presented to the school council (home: this will lead to discussion and curriculum debate).

5. Do nationally prescribed curricula lead to improvement in standards?

Yes, it is reasonable to expect, but depends on definition of the terms.
- What can be expected is more cohesion in the curricula and in the standards.
- No guarantee that results in terms of students' achievement will be better.
- It seems that a balance of a central obligatory curriculum and a system of implementation/innovation/and freedom of schools should have the best effect on standards.

6. Curriculum design

Research needed into:
- how curricula are presented.
- their content.
- core and optional parts.
- criteria for formulating attainment targets.

Research needed into:
- the control of the implementation of curricula, e.g. through detailed course programmes, approval of textbooks, etc.
- the Council of Europe should support especially the Eastern European countries

in developing expertise in curriculum design. This can be done by organizing cooperation between Eastern and Western countries in curriculum development.

REPORT OF GROUP 3

Chairman: Alfred Buhagiar (Malta)
Rapporteur: Charles Mizzi (Malta)

A. Educational practice regarding the inculcation of attitudes and the promotion of values in (written and hidden) curricula

Research and experience seem to indicate:
- The process of formulating objectives and aims for the area of the curriculum has made sufficient headway. The objectives have, however, not always been effectively translated into practice.
- Formal declarations meant specifically to promote values and change behaviour can be ineffective.
- Allocation of specific lessons intended to promote a specific issue (e.g. peace education, human rights education, etc.) can be counter-productive. The approach must necessarily be interdisciplinary and cross-curricular and reflect a commitment of the school as a social organization.
- Strategies and tools for the dissemination, implementation and evaluation of this area of the curriculum have still to be further developed, and research institutions should channel a substantial part of their activity towards resolving the problem. Examples of good practice should be compiled, analyzed and discussed in (ideally school-based) INSET.

Three factors are conducive to a true implementation of the objectives:
(i)   the school environment itself, which should reflect the target attitudes and values;
(ii) an initial and inservice teacher education programme responsive to the topic;
(iii) a proper recognition of the cultural and social context within which a school has to operate and adaptation of approaches.

The European dimension requires attention to be paid to common European history and cultural heritage and this should be compulsory. Some countries would need to give more prominence to social sciences, humanities and modern languages.

- Optional elements in curricula are vital to motivate pupils but educational/ vocational choices should not be imposed at too early a stage of the curriculum.
- The curriculum should include approaches which promote collective experience through social projects.
- The goodwill of different pressure groups (e.g. teachers, parents, trade unions, university students) is essential for the success of educational reforms. If educational reform is to be successful at the level of the classroom, the understanding and support of the pupils is essential.

- The attitude of parents towards innovation must be continually fostered through public relations at both school and national levels.

B. Strategy to be adopted by policymakers contemplating in-depth restructuring of the curriculum

The item is highly topical for Central and Eastern European countries currently in transition from a command society to an open democratic environment.

- The policymakers should recognize the need and use of national expertise in curriculum development and education advice. Such expertise should be supported and encouraged.
- The support of international and European organizations and agencies should be sought in order to provide training opportunities for educational researchers, curriculum developers, teacher trainers, teacher advisers, etc.
- Policymakers should realize that good educational practice is evidenced in various models and that an awareness of the various possibilities is of the essence.
- Research workers should provide decision-makers with critical, analytical accounts or practice in their own and in other countries, paying attention to differences in context.
- Existing networks (EUDISED, INED, IEA, etc.) should involve the institutions and new curriculum development centres in Central and Eastern Europe. CIDREE-East should be set up for interested countries, for example :
  - (i)   establishment of the basic standard which should be achieved by a European citizen.
  - (ii) undertaking joint European research project.

C. Recommendations for further research and development

- New pedagogical approaches designed to help students enjoy the experience of learning and continue learning throughout their lives.
- Possible approaches for counteracting disaffection and dropout due to too abstract a curriculum, knowledge which is too fragmented and delivery which is too dull and repetitive.
- Possible measures for providing adequate opportunities for students who are highly motivated.
- The destabilizing effect that the transition period from one level to another in the educational system has on the learning competences of the child (research has indicated that the negative effect of the transition is appreciable).
- Developing criteria for judging the quality of curricula.
- Investigation of examination design at the end of secondary school education, including the expectation of what should be achieved at this level, the type of examination that should be set and the type of certification that it should lead to, should be discussed at a future workshop.

Part Two

CURRICULUM REFORM IN CENTRAL AND EASTERN EUROPE

César Birzea

Director of the Institute for Educational Sciences, Bucharest

During a seminar dealing with East-West cooperation organized at Loccum in March 1992, Mr Libor Paty, Deputy Minister of Education of Czechoslovakia, stated that the countries of Central and Eastern Europe face a paradoxical situation: they should change a car's engine while travelling by that very car. Within the same framework, Mr Alfred Janowski, State Secretary at the Ministry of Education in Poland, used another suggestive metaphor: in fact, these countries find themselves in the impossible situation of making a genuine live egg out of an omelette.

No matter which of these two metaphors is closer to reality, it is obvious that an important region of Europe is experiencing today a real period of changes, crises and uncertainties known as **the transition period**. It is a unique, dramatic experience to which nobody can offer ready-made solutions. This is a period of rebuilding and reforms, whose duration is impossible to foresee now, but which will surely last for at least one generation. Briefly, it can be characterized by one single sentence: it is the period when a new policy and new objectives should be achieved using the old structures and the same people.

Against the new Eastern European background, the educational problems are so numerous and urgent that even establishing a priority order represents a hard-to-take decision. The public opinion and various pressure groups ask for a rapid change of structures and institutions, curricula and textbooks, finances and administration, methods and means, teachers' training and status. In one word, everything should be changed or **almost everything**.

In a recent study (Heyneman, 1991), an expert of the World Bank said that, in fact, the educational systems of the ex-communist countries represented a machine in working condition, but wrongly used. It was not the machine to blame, but the content processed inside, to wit, the knowledge, skills and competencies required by the authorities. Unlike the underdeveloped countries, Heyneman says, the countries of Central and Eastern Europe need not now build afresh their institutions and minimum infrastructures, training the first teaching staff, elaborating textbooks and instructional technologies. These educational systems have already achieved general schooling until the age of 14, with a low percentage of drop-outs (about 2%) and a satisfactory inner efficiency. They ensure good training in the sciences and mathematics (as I.E.A. comparative studies have pointed out) and, in general, they are provided with minimum resources guaranteed by the state (textbooks, equipment, buildings, salaries). Heyneman comes to the conclusion that the problem of these educational systems is closely linked to content and the manner in which this huge learning machine was perverted by the communist ideology and deeply exploited in the interests of the Party.

It was the **curriculum** that was mostly influenced by this educational policy. This is the reason why the first changes in the transition period have taken place at the level of curricula. We are referring on the one hand to the so-called mending measures aiming at immediately correcting the most striking effects of communist education, and on the other hand to the measures of planned changes integrated in a long-term reform.

The share of each type of measures and their real content obviously vary from one country to another, but generally speaking, we consider that the following objectives are taken into account:

- changing the general outlook on curriculum and defining a new philosophy of education.

- descentralizing the decision-making in the area of curriculum development.

- eliminating the influence of Marxist ideology residing in diffused, transdisciplinary contents, as well as in separate school subjects taught exclusively for political indoctrination.

- assuring a greater importance to humanities, foreign languages, social sciences and philosophy.

- replacing the rigid bureaucratic control by a system of formative evaluation integrated within the educational process.

- forming and training new competencies, non-existent or inhibited by the old power, but absolutely necessary for living in a democratic society and under the circumstances of the market economy.

We shall now deal with these six common tendencies.

1. **Curriculum** was a term seldom used in the countries of Eastern Europe and the ex-Soviet republics. More likely than not, when Western texts about the curriculum had to be translated, periphrases or notions were resorted to covering only certain peculiar aspects: syllabus, course, content, instruction, education, teaching activity, etc. Even today this term is not known by the public at large and when it is used, there is still a lot of confusion. It is true such confusion does also exist in Western specialized literature, where a single author (Novak, 1960) could identify 98 different uses of the word "curriculum" in the pedagogical texts written in English alone. It is also true that in the Latin countries of Europe or in those with centralized

administration the concept of curriculum is a recent acquisition, being assimilated only during the last 10 years. But what is most obvious is the fact that in the final analysis it is not naming that matters, but the philosophy involved. From this viewpoint it seems to us the countries of Central and Eastern Europe, as well as the ex-Soviet republics, show an ever greater interest in the democratic implications of the term "curriculum".

In fact, it is well known that the philosophy of the curriculum shifts the stress from bureaucratic documents to the decision-making preceding, going together with and following official acts. Consequently, the main interest is no longer focused on the lists of compulsory subjects, but on the manner in which the teaching, learning and evaluation situations are being organized. Briefly, instead of rigid documents and formal regulations, **operational curricula** are required.

Traditionally, the lists of knowledge content were worked out at the centre by a restricted group of initiates, who were rigorously checking ideological correctness and content compatibility with the power interests.

The products of these supercentralized decisions were the rigid prescriptions and exclusive textbooks officially enforced by authority. But it was much more: these instructions given by the centre permitted only a certain interpretation of school syllabi, namely the official variant resulting from the state ideology. To be sure of it, the leading power established a large control body (inspectors, political activists, informers), who pointed out and severely punished any deviation from the "good method", from the "right" or "scientific path" approved by the Party.

This situation is changing. The first rectification measures of the transition period were in fact authoritative changes too, but in an opposite direction: the opponents of Marxist ideology and of the one party system have replaced by central decisions the contents and curricula of communist education. Obviously, these decisions met the demands of the people at large, but they were still taken within certain centralized and authoritative structures.

What interests us in a long-term perspective is the very change of the decision-taking process. "Curriculum" is not merely an exotic term or a new pedagogical fashion, but a liberal conception, which looks upon any educational programme from a triple perspective: as a **project**, it expresses an ideal or an option of educational policy; as a **normative text** it aims at adjusting the educative relations and conditions in a given context (teaching a certain subject-matter for a certain period of time and in a certain school environment); finally, as an **action guide** defined by each teacher in his turn, it contains the immediate guidelines of the educative situation. At this operational level, as exact answers as possible are given to the following questions: **who** (the subject of the action), **why** (the educational objectives), **what** (the contents), **how** (the instructional strategies) and **to what extent** (evaluation of results). This highlights the difference between the traditional lists of contents, exclusively limited to the **what** question and the broader, more accurate approach of the curriculum.

Of course, the transition to a new outlook on the curriculum will take time. It requires a comprehensive programme for staff training, as well as renegotiating responsibilities at all levels: national, central, regional, local, and institutional.

2. This analysis points to the absolutely necessary condition of a more liberal curriculum practice: **decentralizing the decision-making process**. Actually, even after the recent changes, school curricula are still elaborated at the centre, in ministries or in governmental offices by a limited group of experts. This fact has two possible reasons: on the one hand, the state budget cannot go to new expenses on alternative curricula and textbooks during this period of financial restrictions. Generally, in the countries of Central and Eastern Europe the education budget represents less than 10% of the gross national product; out of this, over 75% is allocated to salaries and this percentage tends to increase because of uncontrolled inflation. On the other hand, exclusive top-down curricula are still preferred, because most teachers and inspectors are not acquainted with curriculum methodology yet. Nevertheless, we can already notice new tendencies favourable to decentralized curricula. More precisely, there is a transition from a rigid, authoritative planning to a more flexible central coordination. Thus, in Poland, Romania, Bulgaria and the Ukraine, curricula continue to be coordinated by education ministries, but local adaptations of official curricula are allowed as experimental approaches. In these countries, where there are central departments for curriculum design, there have been created also relatively autonomous commissions and working groups made up of inspectors, researchers and teachers.

An even greater decentralization has occurred in Hungary, Czechoslovakia and Byelorussia, where certain decisions concerning curricula have been transferred to the level of districts and school units. So, for example in Byelorussia, municipal secondary schools financed by local authorities have been set up; they can establish their own curricula and are also free to employ teachers by direct labour contracts. In Czechoslovakia, the new regulations stipulate that 20% of teaching activity should be organized by each and every school as it desires, according to local demands and interest. Generally speaking, in the countries of Central and Eastern Europe, as well as in the ex-Soviet republics the number of compulsory subject areas was reduced, increasing the number of the optional ones instead. In Poland, for instance, the weekly schedule was reduced by 10%. In Hungary, out of 32 weekly lessons in upper secondary education, optional subjects represent 19%. In Romania, the optional courses cover 10% of the weekly lessons taught in upper secondary education, while in Bulgaria their percentage varies between 6-12% according to grades.

There is another important question linked to these new developments. Are these countries already in a position to pass to free competition adopting market laws in the field of curricula and textbooks or should they go on subsidizing educational publications, thus accepting the well-known disadvantage of any state-paid industry ? It is only in Czechoslovakia that the state monopoly on school textbooks has been abolished and competitive textbooks already admitted (cf. Walterowa, 1991). As to the other countries, the liberalization of the educational market is an objective put forth by Hungary, Romania, Poland and Bulgaria but only for the year 1995. For the time being, the production of textbooks is carried out by state-owned publishing houses. Even state-paid, textbook production is influenced by the liberalization occurring in the book industry: the price of paper has become prohibitive, continuously increasing the purchasing cost of textbooks and the printing-houses often prefer more commercial titles. Moreover, this situation renders almost impossible the manufacturing of other printed educational materials (tests, pupils' workbooks, text or problem collections, laboratory copybooks), absolutely necessary for diversifying curricula.

3. What is really obvious, irrespective of the pace of all these reforms, is the fact that **communist ideology** is no longer the official ideology of curricula. In fact, the first rectification measures of the transition period were just aimed at immediately eliminating the subject areas of political indoctrination, as well as the ideological contents deeply disseminated in all the other school subjects. These measures occurred in the middle of the school-years, the pupils having started courses with the old textbooks of history, geography, political economy, scientific socialism, etc. It was an abrupt change, embarrassing for certain teachers; during some weeks the pupils suddenly learnt that their history textbooks contain many mystifications and in fact, Marxism is not at all an unfailing, universal doctrine.

But in the new curricula new subjects have been reintroduced, which had no place before, because of their possible competition with official ideology. This is the case with philosophy, economics, aesthetics, ethics, psychology, logic, sociology, civics. For certain subjects (history, geography, literature) some chapters were excluded even during the school-term; others were added by simple juxtaposition, but preserving the conception they had been initially grounded on.

This situation makes more necessary and urgent an essential reform of the curriculum and a new outlook on teaching social sciences and the humanities.

4. As a matter of fact, the whole body of knowledge, skills and competencies is under revision. Traditionally, according to the materialistic and utilitarian philosophy of communist education, the humanities, social sciences and foreign languages were of minor importance. On the contrary, great attention was paid to productive work, meaning in fact the pupils' and students' forced labour; its targets were more economic and ideological than educational.

The new curricula not only gave up the courses of political indoctrination and paramilitary training, at the same time reducing and redefining practical instruction, but they have also increased the number of lessons for **the humanities, social**

112

sciences and **foreign languages**. In most of the countries under discussion, the study of the first foreign language begins in the first grades of primary education; the teaching/learning of the second foreign language starts in the 5th grade. A third foreign language may be studied in the high school, too. If before 1989 Russian was the main foreign language in Eastern curricula, now there is a shift to English, French and German. Thus in Poland, Czechoslovakia and Albania many of the former teachers of Russian cannot go on with their job.

Generally speaking, in conditions of great linguistic and ethnical diversity there are hot debates on the relations between the official language, the language of instruction and the mother tongue. These disputes have a direct bearing on curricula and the educational systems are often called on quickly and equitably to solve social and national problems brought about by the respective region's long, tumultuous history.

It is enough to give one single example. In the Ukraine, where more than 27% of pupils speak at home a language different from Ukrainian, one of the first measures was to give the right of using one's mother tongue in primary education to over 110 nationalities living in this country. At the same time, in secondary education, Ukrainian covers 19% of the weekly educational activity, just to encourage the study of the new national language.

Finally, in some of the ex-Soviet republics (especially Baltic countries and Moldavia), an even more difficult problem has arisen. The switch-over to the Latin alphabet has meant that these countries have had not only to revise curricula, but also suddenly to change all textbooks and school documents. We can easily imagine the great sacrifices involved by such a decision taken in a period of severe economic austerity.

5. Another bone of contention relates to **evaluation**. At first sight, exams seem to be similar all over the world, being a technical problem, independent of philosophical options or ruling ideology. In reality the situation is quite different, because far from being a mere instrumental choice, evaluation methodology reveals a certain pedagogical conception, a general attitude towards the place of control in social life. In authoritative educational systems, the main purpose of control is to penalize, to classify and select. Consequently, they give priority to final examinations, which find, record and punish learning mistakes. The situation is different in the educational systems grounded on a liberal outlook. They consider evaluation as an improving measure, as a continuous correction, integrated in the education process itself. In this manner, evaluation acquires a formative character; it becomes meliorative and re-educative. It intervenes promptly at the very learning moment and after each curriculum unit (lesson, chapter, module). At this micro-pedagogical level, the learning errors can be still remedied before leading to school failure.

This new outlook on evaluation is gradually gaining ground in most of the countries included in this study. Thus, in Poland and Hungary a new system has already been tried out; it focuses more attention on formative evaluation than on the final assessment. In Bulgaria, experts are working hard now to define the minimum standards of knowledge essential to any criterion-referenced evaluation. And in Romania, as well as in Czechoslovakia, the exams at the end of secondary education (the "Baccalaureat" and the Maturita' respectively) have been under revision. Both pupils and their parents have questioned their relevance.

In conclusion, we can say that irrespective of the changeable configuration of political and economic developments, it is obvious there is no way backwards and today's generation of pupils and students will live in another type of society than the one they were born in. But this generation needs a new type of training from this very moment. More precisely, **proper competencies** should be formed for living in a democratic society and working under the conditions of a market economy. This implies more than just content changes. It involves long-term action, more difficult indeed, because it deals with mentalities, structures, attitudes, social relations and institutions. The new curricula try to harmonize this long-term approach with immediate corrective measures. They have included knowledge regarding the market economy, management and finances, at the same time encouraging the enterprising spirit to a greater extent. They emphasize quality, aiming at achieving a new balance between general culture and professional training. These curricula include new contents and new subject-matter covering human rights, democratic institutions and principles, man and the environment, the quality of life, current social problems, mankind's everlasting values. The next step will be to turn these contents into living abilities, to translate knowledge into operational competencies and diminish the gap between projects and daily reality. Without this outstanding step forward, the whole system of promised reforms remains at the level of rhetoric or within the framework of political speculation.

REFERENCES

Birzea, C. (1991) *The Educational Policy in the Transition Period: the Case of Romania,* Oslo, International Seminar on the Situation of Education in Central/ Eastern Europe and the Soviet Union, November.

Darvas, P. (1991) *Perspectives of Educational Reform in Hungary.* In: M.B. Ginsburg (ed.) *Understanding Educational Reform in Global Context, Economy, Ideology and the State,* New York, Garland Publishing.

Développement de l' éducation en Bulgarie (1990), Genève, Conférence Internationale de l'Éducation.

Développement de l' instruction publique dans la R.S.S. d' Ukraine (1990), Genève, Conférence Internationale de l'Éducation.

The Development of Education within 1989-1990, Warsaw, (1990) Ministry of National Education.

Éducation nationale en R.S.S. de Biélorussie, Minsk, (1990) Ministére de l'Éducation Nationale.

Hajdaraga, L. *The Development of the Educational System in Albania in the*

Framework of Social and Economic Changes, (1991) Oslo, International Seminar on the Situation of Education in Central/Eastern Europe and the Soviet Union, November.

Heyneman, *S.P.* (1991) *Revolution in the East: The Educational Lessons*, Oxford, Oxford International Roundtable on Educational Policy, September.

Kozma, T. (1991) *Recherches sur l'enseignement secondaire en Hongrie*, Strasbourg, Conseil de l'Europe.

Nagy, J., Szebenyi, P. (1990) *Curriculum Policy in Hungary*, Budapest, Hungarian Institute for Educational Research.

Novak, B. (1960) *98 Definitions of Curriculum. The Clearing House*, vol. 34, p. 358-360.

Rassekh, S., Vaideanu (1987), G. *Les contenus de l'enseignement*, Paris, UNESCO.

Vlasceanu, L. (1992)*Trends, Developments, and Needs of the Higher Education Systems of the Central and Eastern European Countries*, Bucharest, CEPES.

Walterowa, E. (1991) *Curriculum Development in Czechoslovakia: Problems, Trends, Politics*, Lyngby (Denmark), Scandinavian Information Seminar IMTEC, October .

REPORT ON THE STATE OF THE ART OF AN ONGOING RESEARCH PROJECT INTO TEACHING ENGLISH AS A FOREIGN LANGUAGE IN SECONDARY SCHOOLS IN AUSTRIA

Herbert Puchta
Pädagogische Akademie des Bundes, Graz

BACKGROUND

School systems and curricular outlines

In Austria, when children leave primary school at the age of ten, they can choose between "Gymnasium" (grammar school) and "Hauptschule" (secondary modern school). The Gymnasium lasts for eight years and finishes with a final exam ("Matura") that enables students to go to university. The Hauptschule lasts for four years and aims at providing the learner with basic general education depending on their interests and abilities and preparing them to enter the world of work. This school lays the foundations that enable students to attend middle or higher school education.

For this reason curricula for the Gymnasium and the first streams of the Hauptschule are identical. After four years at the Gymnasium, students who do not want to stay on have the same choice as the ones from the Hauptschule and can go on to middle or higher schools. Higher schools last for four or five years depending on the type of school. A final exam (Matura) enables students to go on to university. There are various types of Higher Schools. One type (Bundesoberstufenrealgymnasium) offers a more general training which is comparable to education offered at the upper secondary classes at the Gymnasium. Other types offer a more specialized training with emphasis on certain subjects. There are three main types of schools: "Hoehere Technische Lehranstalten" (Technical high schools), "Handelsakademien" (Commercial high schools) and "Hoehere Schulen für wirtschaftliche Berufe" (Economics high schools).

DESCRIPTION OF THE RESEARCH PROJECT

The research project described in this paper focuses on problems that occur in English as a Foreign Language for the students who change over to the various types of secondary schools described above after leaving Hauptschule or after the fourth year at the Gymnasium. The project is being carried out on behalf of the Austrian

Ministry of Education at the Pädagogische Akademie des Bundes in Graz. The two main researchers are Günter Gernrob and myself.

Aims of the research project

The aims of the research project are:
- to analyze the factors that make changing over from lower secondary to upper secondary level in English as a foreign language difficult for the students
- to analyze how the factors that make it difficult to change from lower secondary to upper secondary level in English as a foreign language could be minimized
- to suggest concrete action that could be taken to enable students to change over from lower to upper secondary level with fewer problems.

Duration of the research project

The project is designed to be carried out within a span of two years and is presently in its second year. So far two reports on the state of the art of the project have been presented to the Austrian Ministry of Education. The final report is due in July 1993.

Research hypotheses

As the basis of the research project the following hypotheses were defined:

the curricula for teaching English as a foreign language for upper secondary schools are not sufficiently coordinated with those for teaching English as a foreign language at the lower secondary level.

- practical classroom work in English as a foreign language at the Hauptschule, on the one hand, and at the upper secondary schools, on the other, differs with regard to the teaching methods and especially as far as testing methods are concerned.

- teachers at the various types of schools concerned do not have sufficient knowledge of the curricula of the other recognized schools.

- the opinions held by Hauptschule teachers of the expectations of upper secondary school teachers concerning what students should know do not match what these teachers really expect.

- these false expectations of what is required at the upper secondary schools influence the work done in Hauptschule classrooms.

- the fact that there are no standardized tests to determine the students' level after lower secondary education adds to the problems with changing from lower to upper secondary education.

Research methods

The following methods are being used in the research project:
- analysis of the various curricula
- comparison of the curricula with regard to gaps concerning their coordination
- evaluation of the curricula from the point of view of modern language teaching research and methodology
- empirical evaluation of the practice of testing at the various types of schools concerned
- questionnaires handed out to teachers of the various types of schools in order to analyse teachers' subjective theories (see hypotheses above).

State of the art of the research project

The following phases have been carried out so far:
- analysis and evaluation of the curricula of the various types of schools
- design and distribution of the questionnaires
- analysis of the results of the questionnaires
- collection of tests from the various schools
- analysis of these tests.

A BRIEF SUMMARY OF THE RESEARCH FINDINGS SO FAR

Analysis and evaluation of the curricula

The hypothesis that the curricula for teaching English as a foreign language for upper secondary schools are not sufficiently coordinated with those for teaching English as a foreign language at the lower secondary level cannot be generally verified. The analysis of the different curricula show that no major problems are to be expected with students changing from lower secondary level to two types of schools (Oberstufenrealgymnasium and Höhere Lehranstalten für Wirtschaftliche Berufe). The curricula for the two other types of upper secondary schools, however, show different conceptual views of language learning.

The curriculum for the lower secondary schools stresses the functional role of grammar in language learning in accordance with widely accepted findings in language acquisition research. The curricula of the Handelsakademie and the Höhere Technische Lehranstalten, however, do not elaborate on functional aspects of grammar, and allow for a formal interpretation of the role of grammar in the language learning process. Additionally, in the curriculum for the Handelsakademien, translation is mentioned as an explicit method of practising grammar which is in sharp contrast to current research findings in the area of foreign language teaching.

Summary of the analysis

The following is a summary of the analysis of the tests and of the results of the questionnaires handed out to the teachers at the various types of schools.

The analysis of the tests and the results from the questionnaires from the various types of schools seem to show some decisive factors that make changing over from lower secondary schools to upper secondary level in English as a foreign language difficult for the students.

The study clearly shows a lack of information from teachers of the upper secondary level concerning the curriculum of the lower secondary schools. This might be the reason, for example, that students are frequently expected to translate from German into English in the tests in the first year of three of the four main types of upper secondary schools. This must be a major problem for students who change from lower secondary to upper secondary schools because in lower secondary schools the curriculum explicitly forbids translation as a means of testing the students' language competence and therefore most students are not at all familiar with the specific skills they would need to be able to translate.

In this context it is interesting to note, though, that a small percentage of the tests written at the lower level schools does contain translation tasks although forbidden by the curriculum. This seems to support the hypothesis that there is a certain wash back effect from the opinions of the teachers at the lower level schools of the expectations of the teachers of upper secondary schools concerning what students coming up from the lower level classes should know.

Accordingly, the practice of testing the students' language competence shows gross differences in aims and methods between three of the four types of upper secondary schools on the one hand and lower secondary schools on the other. In these upper secondary schools, tests show an absence of communicative tasks. Receptive skills (reading and listening comprehension) hardly get tested at all.

These findings show a remarkable contrast to the subjective theories teachers expressed in the questionnaires. In all the various types of upper secondary schools oral communication was regarded as the most important area of language teaching. However, when asked about the most frequent source of dissatisfaction with the students' competence when they entered the first year of the upper secondary level, teachers of these upper secondary schools complained most frequently about a lack of knowledge of grammar.

EDUCATIONAL RESEARCH IN CYPRUS AND REFORMS IN SECONDARY SCHOOL CURRICULUM

Christos Georgiades
Inspector, Ministry of Education, CY - Nicosia

Educational research in Cyprus, mainly carried out by the Directorate of Secondary Education and the Pedagogical Institute, aims at providing data and assistance to the Ministry of Education, necessary for its developmental function. Research creates a bank of scientifically supported information to be used by the educational authorities in decision making, as well as by teachers in organizing their teaching activities.

The Research projects fall under the following main categories:
- developmental research and evaluation of project work,
- comparative studies and achievement norms in various subjects,
- evaluation of textbooks and teaching material,
- evaluation of curriculum content,
- needs for teacher in-service education,
- pupils and teachers characteristics,
- the role of parents and social factors,
- school and work.

The significant results and findings of the research carried out are utilized by the Ministry either for remedial action or for major educational reforms and innovations. Based on the results of educational research carried out in recent years, the following new policy orientations and priorities have been decided:

- the setting up of a University,
- improvement of the quality of secondary education,
- the setting-up of an integrated nine-year educational programme and restructuring of the educational system,
- increase of the proportion of pupils attending technical and vocational education to 25% of the pupil population,
- integration of education with the world of work and life in general,
- development and upgrading of the skills of adult workers with a view to enabling them to find a job, meet the needs of modernization in work conditions and increase productivity.

In the case of Secondary Education, research findings lead to an intensification of in-service training courses, while greater importance has been given to the content and quality of education. Improvement of quality is pursued by:

- modernizing syllabi and preparing new teaching material,
- introducing educational programmes at various levels aiming at raising standards and reducing the percentage of school failure in general and functional illiteracy in particular,
- introducing the subject coordinator for all subjects taught in secondary schools on an equal footing with the existing institution of the Deputy Head,
- linking secondary schools more closely with the world of work in an effort to see education as a preparation for life.

However, the most important reforms and innovations at secondary level have been (a) the introduction of the 9-year integrated educational programme and more recently, (b) the reorganization of the system of options and (c) the introduction of Health Education.

Introduction of the integrated educational programme. Research has clearly shown that the existence of a gap between primary and secondary level is a reality. Both from the point of view of philosophy, methodology and priorities and from the point of view of expectations there is a remarkable difference between the two levels - though both primary and secondary education are state-controlled. Secondary school teachers, experts in their fields and all university graduates know very little about what the priorities are and what methodologies and strategies are used by primary school teachers who are not specialized in a given discipline.

This has been clearly shown by research and observation. Thus the decision has been taken to link and to continuously coordinate schooling from age six to the completion of the first 3-year cycle of secondary level (age 15). This innovation is under continuous observation and study for a more effective organization.

Reorganization of the system of options. The system of options, originally introduced during 1977-78, proved to be heavily academically oriented, offering mainly theoretical knowledge divorced from the realities of everyday life without giving the opportunity, mainly to the low-achievers, to choose individual subjects necessary for their preparation for life. Indeed, instead of giving the opportunity to choose any subject, it became a rigid system of opting for compilations or sets of subjects and inevitably the student finds himself facing a subject within the combination, he is not really interested in. The newly proposed system provides for a common core course at form A which affords this form a period of familiarization and orientation within the lyceum (upper cycle). This will contribute towards avoiding early or immature decisions. At the same time, the option of sets or combinations of subjects is abolished and the student is now free to choose any subject he is really interested in.

Health education. Equally important is the decision to introduce a spirally organized Health Education course founded on the basic health needs of society. Recent health issues, mainly AIDS, drugs, heavy smoking, sexual abuse, alcohol, road safety and heart diseases, to name a few, and data gathered by research studies convinced

government to approve and support a programme of Health Education which is based on three foundation stones of teaching and learning:
- personal development
- social development
- the community and the environment.

Research findings in Cyprus are submitted to the Council of Europe. The executive instrument in Cyprus for keeping EUDISED (European Documentation and Information System of Education) informed of research project findings in Cyprus is the Pedagogical Institute. The information given is processed through the Computer Research Service of the Council of Europe and it is made accessible to all countries of the Council of Europe through the E.S.A. (European Space Agency).

SECONDARY SCHOOL CURRICULUM IN THE FRAMEWORK OF THE CURRENT EDUCATIONAL REFORM IN CZECHOSLOVAKIA

Eliska Walterova

Institute of Educational and Psychological Research
Charles University, Prague

INTRODUCTION

Problems of curriculum development are among the most complex and crucial in any educational system. They are affected by a wide range of factors - political, social, economic, as well as by national tradition, culture and context.

It we go deeper to the substance of curriculum we can see that the concept of man and the responsive concept of education are the philosophical stratum of curriculum. The curriculum is determined by a hierarchy of values, by the functions of education and the school as it is perceived by the community as well as by the character of the environment and by the human qualities of individuals involved in the educational process. The curriculum depends on the quality of social experience as well as on the ways by which it is transformed.

Concept of Curriculum

The concept of curriculum has changed during the last decades. The older narrow concept concerned the selection and sequence of the content answering the question WHAT to teach. Curriculum reforms in the 1960s in many countries initiated research on social needs and projected rational goals. The broader concept of curriculum involves the total programme of educational institutions and answers the questions WHOM, WHAT and HOW to teach (D.K. Wheeler, 1974). Consequently significant discrepancies between curriculum as a project (plan) and its performance in school were identified. The phenomenon of the "hidden curriculum" was also observed and became important. Research on instruction and on the evaluation of educational effects of the hidden curriculum contributed to a more precise interpretation of this phenomenon.

Present Concept

The present concept of curriculum involves not only content and the planned level of knowledge transmission: curriculum refers to the educational context, the structure of the teaching-learning process, and, more generally, to "every experience that a child undergoes during schooling and the educational aims and goals, courses, classroom activities, staff-student relationships. It also covers the reaction between teachers, resources and many other factors, which impinge on the teaching - learning situation in schools" (OECD, 1983, p. 59). I define curriculum in brief as a life course of learners in an educational institution (E. Walterova, 1991 b). This concept of curriculum offers a plastic, even holistic, reflection and presents the hidden curriculum as a transparent one. If we want to operationalize this concept of curriculum we need a broader paradigm than the one that raises the questions of whom, what and how to teach only. We have to involve the following questions as components of education:

WHY?	sense, values, functions
WHOM?	learners' characteristics
WHAT?	content
WHEN?	time factor
HOW?	learning strategies
UNDER WHICH CONDITIONS?	learning environment
WITH WHAT EFFECTS?	expected results

Answers to these questions are interrelated and should be concretized and developed in more detail at every level: supranational, national, local and class, as well as in every educational situation. Decisions concerning all these curriculum questions are the problems of current educational reforms in many countries. These reforms should be more qualitative in their character than previous ones. They should be aimed at a deep democratization and humanization of education. They demand changes at the macro-level of education as well as at the micro-level of instruction in schools (W. Mitter 1991).

Curriculum development in Czechoslovakia has to undergo substantial change as has the whole of the education system. Such change is considered, in fact, as one of the social priorities for the country.

The task of democratization and humanization of education is particularly demanding in Czechoslovakia and in the other formerly socialist countries of Eastern and Central Europe. Even in these countries "a crisis of the absolute horizon" (V. Havel 1990) has led toward a decomposition of horizons and toward the disintegration of man into his anonymous roles of consumer, patient, elector and pupil, toward the devastation of human values, nature, and the environment.

The main purpose of this paper is to discuss curriculum development in Czechoslovakia and to clarify problems in the framework of current educational reform.

The main problems connected with the fulfilment of this task have been:

a. to find an appropriate concept of curriculum: the concept corresponding with a broad paradigm (see above) is considered to be most appropriate for a complex understanding of the problems.

b. to describe the present state because the process of transition is very dynamic. Recent trends and characteristics of curriculum are starting points for the explanation. Educational reform is clarified by its strategy, priorities and considerations about its possible trends.

c. to find a balance between general questions and trends of curriculum development, including those found in other countries, and specific ones needing a comparative approach which in turn leads to consideration of the needs of comparative research on curriculum development in an international context.

PROBLEMS OF CURRICULUM DEVELOPMENT IN CZECHO-SLOVAKIA

What are the problems connected with the transformative process of curriculum development ? To understand them one must know more about recent trends in curriculum development.

Recent trends

In 1948 the Act on comprehensive schooling was passed. After that the Czechoslovak educational system underwent several reforms and many changes. They were, unfortunately, stimulated much more by political and administrative decisions than by a reasonable search for the optimal and appropriate model.

The former progressive concept of comprehensive school (which was discussed even in the 1930s) was gradually deformed. The school system was made rigidly uniform. It no longer allowed support for different learners' potentialities, beliefs, interests, orientations. An ideological framework of curriculum was enforced by the utilitarian philosophy of Marxism-Leninism, communist morality, collectivism, the class struggle and a disciplined subordinated concept of man. It was intolerant of any alternative thinking and feeling, particularly if of a religous nature.

The educational track was linear and the curriculum was very closed. Systematic evaluation of effects and outcomes of education as a feedback from schools was missing.

The social credibility and cultural status of education fell even further, particularly since the last educational reform in 1976. Then the most regressive measures were enforced. Consequently the divergence from curriculum trends in other developed countries was reinforced:

a) The curriculum for the school population up to 14 years (pupils in the comprehensive, 8-year basic school sector) was strongly unified,

b) The curricula for the three streams of secondary education (general, technical, and vocational) were fused together. This led to a paradoxical situation: technically and practically oriented learners of vocational schools were forced to study compulsory academic subjects in a theoretical setting.

At the same time secondary general school (gymnasium) students were forced to attend technical courses in spite of their university orientation. Neither were the academic courses (those were reduced to accommodate the technical subjects - see appendix 3) nor the professional training efficient. The credibility of secondary general education was particularly degraded.

Curriculum policy

Curriculum policy in Czechoslovakia was strongly centralized and authoritative. Not only were key decisions about the structure of the educational system and the concept of its grades and types of school made at the central (State) level, curriculum decisions for every grade and type of school, syllabus, textbook and teacher manual were also made centrally. Central curriculum projects and materials were taken as a norm for every school and their compulsory fulfilment was prescribed (even the time spent on every topic was planned). The absence of curriculum development at the middle level of school produced many problems and negative consequences. Local needs and the social background of schools and their populations were not taken into account. The limited participation allowed to teachers meant that they felt both dependent and constrained . Some of the teachers modified the central curriculum with feelings of uncertainty and uneasiness. They were thus limited in their professional freedom, independence, and creativity.

Characteristics of curriculum

This uniform, directly manipulative curriculum was strongly content-centred and discipline-oriented in its design as well as in its performance.

The task of designing the curriculum was delegated to *ad hoc* groups of selected designers. They were mostly scholars and school administrators, only rarely teachers, and never psychologists. Their activities were focused mostly on the selection of (better described as "bargaining" over) the instructional content.

126

Traditional science disciplines and their structures were preferred as a source of this content. Some disciplines and topics (philosophy, psychology, anthropology, aesthetics were largely neglected) and practical courses (family life, medical care, home economics, etc.) were missing. The curriculum as a whole was over-crowded with theoretical, abstract, very often dead-alive knowledge and numerous items of rigid information.

The verbal and encyclopaedic character of knowledge prevailed. Learners' beliefs, experiences, attitudes and emotions, also their age-specific needs were very often neglected. These qualities of content and the emphasis on rote learning resulted in cursory knowledge, and the passivity, dependency and mediocrity of the learners.

The instructional process in which frontal learning methods prevailed as well as the current system of examinations and questioning of pupils, the absence of dialogue between teachers and learners, the support of memorization instead of problem solving, and the lack of development of thinking skills, led very often to frustration and stress for the learners, or to their indifference. This being the case at the current time what is evidently necessary is a critical revision, even reconceptualisation of the curriculum.

EDUCATIONAL REFORM AS A FRAMEWORK FOR CURRICULUM DEVELOPMENT

The problems described above should be solved not only by additive innovations or by changes in curriculum design, but also through broader and deeper educational reform, to be realized gradually, and to be supported (even initiated) "from the bottom up" by teachers and to be accepted by parents and learners.

First steps of transition

The situation with regard to education in Czechoslovakia has changed as a result of major changes in Czechoslovak society as a whole.

Shortly after November 1989 immediate steps regarding the curriculum were taken:

- at all levels of schooling the teaching of Marxist doctrine was abolished and different subjects (the social sciences, humanities, aesthetics and religion) were offered instead.

- textbooks on civics and history (and some topics from other subjects, such as geography and literature) were removed from schools. The reason was their ideological orientation and falsification of history and some other issues. Consequently the absence of textbooks has created problems for students and teachers.

- new textbooks are being developed. The state monopoly on textbooks has been abandoned and alternative textbooks are now allowed or recommended.

- the teaching of foreign languages was changed radically. Instead of Russian which was compulsory in all types of schools (and for about 60% of the population the only foreign language learned at school) students can now choose from English, French, German or Russian and, in some schools, Spanish.

Serious practical problems resulting from this change are a need for requalification of the Russian language teachers and a lack of qualified teachers in the other languages.

Diversification of education

According to the Proposal for Innovation of the System of Basic and Secondary Schools (1990) the structure of the educational system (confirmed by the previous Law of 1983) was changed and diversified (compare appendix 1 and 2):

- besides state schools, private and church schools can now also be established and the number of these is growing rapidly. They develop their own alternative, original concepts of curriculum. They could also inspire and support a more creative climate and competitive atmosphere in other schools.

- the disengagement of the educational system from a unified structure allows different alternative types of secondary schools to be established, (8 and 6-year gymnasia from the age of 11/12) or specialized schools stressing languages (modern or classic), mathematics, gymnastics and sport, schools which teach some subjects in a foreign language etc. (Table of weekly lessons: see appendices 4 and 5).

- according to the new system of school-based management (since 1990) schools are more free to make decisions concerning the curriculum. About 20% of the teaching time is left to each school for decision-making on what is to be taught.

Trends of changes

Education should now feature as one of the political priorities in Czechoslovakia. The relaxation of the social and political atmosphere in CSFR has created more appropriate conditions for modifications, alternatives and experiments to take place. Alert schools and creative teachers could change their own concrete situation and be engaged in the reconceptualization of education and discussion on educational reform.

The next part of this paper will refer more specifically to the situation in the Czech Republic, which is, in general, similar to that in the Slovak Republic, although in some specific features of the system slight differences exist.

Different initiatives aimed at educational reform in the Czech Republic led to five alternative proposals made by

- an expert team headed by the Dean of the Faculty of Education, Charles University.

- an educators' and teachers' initiative team,

- a team of mathematics and science teachers,

- the ministry of Education Commissions and

- the market economy orientated group IDEA.

These proposals have been published and should be discussed publicly in the immediate future.

Strategy of reform

Recent authoritative, administratively prepared, and implemented reforms were unsuccessful, leading to public dissatisfaction because they harmed both teachers and learners. The strategy of the present reforms should be to learn from the negative experience of previous reforms and to avoid it. Next, popular support is basic for their success:
- reform must be developmental. It should undertake the appropriate steps gradually.

- reform should be accompanied by legislative measures establishing the conditions for its performance.

- reform should aim at increasing general educational standards, it should receive political and social priority as well as aim to develop the particular personal capital of each individual.

- standards of education should be high corresponding to those in developed countries, particularly European ones. This is an important point for future integrative processes within the European structure where Czechoslovakia wants to participate.

Priorities of educational reform

The next set of principal questions takes account of priorities classified as follows:

- *Humanization of education.* How can an independent, responsible and tolerant personality be cultivated ? How can human values, creativity, cooperation, synergy and meta-cognitive competence be supported? How can the climate of schools be changed to become more humane?

- *Democratization of education*. How can we establish conditions for respecting the needs and demands of different groups and individuals including compensation for handicaps ? How can we support the optimal development of every learner when individual potentials are different?

- *Pluralism in education*. How do we reshape the structure of the educational system? How do we achieve pluralistic options on the educative track? How do we diversify education, looking for pathways for learning in every period of life?

Curriculum Development Priorities

Aims and functions. An increase in the general educational standard of the Czech population, support for the general cultural base of the curriculum, and harmonization of objective and subjective needs are priority aims. Polemical debates reflect controversy between economically pragmatical concepts (supporting earlier specialization and professionalization) and broader general humanistic concepts (postponing the decision concerning the learner's career and specialization to an older age).

The learner as the subject of education. To change the authoritative approach and content-centred orientation of the curriculum into a curriculum which is liberal and pupil-centred is a serious and difficult task. The learner's development is considered an important starting point for educational reform. His right to respect and partnership in the educational process should be supported. Most views are coincident at this point, however the heritage of the recent authoritarian past will continue to influence the human factor and practical life in the schools, and the period of transition will undoubtedly continue for a long time before it is completed.

Content reconstruction. The concept of educational content should take into account the local environment and national and ethnic specifics. But, beside the national it should also include the international dimension (particularly the European dimension). Trans-cultural, global and universal dimensions of content should be of special interest to curriculum planners. Disharmony in the debates on content concerns controversy over two contending points of view, namely : whether to develop a disciplinary approach which prefers the science field as a main source of content and defends traditional subjects, or an integrative approach in which content is supposed to be projected as a whole and then distributed into subjects.

According to the second approach the main components of the content - core curriculum - should be expressed in such categories as communication, information processing, independent thinking, problem solving, human understanding, nature, technology, a sense of science and self-development.

130

In any case, content should be oriented more to meta-cognitive knowledge and to core knowledge. The quantity of learning material has to be decreased. The organization of the content in instruction should offer more integration, flexibility, non-traditional topics, courses or subjects (at least as optional ones). The number of optional lessons should be increased and the number of common core lessons should be decreased. The standardization of content and the necessity of objective evaluation are problems to be discussed. But the question remains how to solve the discrepancy between the demands of curriculum liberalization and the necessity for the standardization of the educational output of schools. It is about the differentiation between two levels of knowledge acquisition, the variants being general practical and special theoretical.

The progressive tendency is to establish a "minimal" core curriculum in every subject. It is being supported particularly by teachers and textbooks designers.

School environment. Many considerations and visions of the method and organization of instruction have been discussed as well as proposals about how to change and improve the atmosphere and school environment. In this respect a long transformative process is expected with the initiatives for change coming, ideally, from the schools and teachers themselves.

Evaluation. One very substantial curriculum problem that needs urgent attention is the assessment of learners and the evaluation of education outputs. No objective criteria or system of evaluation currently exists. Czechoslovakia has not been involved in international projects about evaluation involving particularly the IEA. In schools it is common to question and test learners frequently and give them marks on all their work. It is not exceptional for a pupil in basic school to receive 5 or 6 marks in one day. These marks rarely have the function of diagnosis or feedback. They are, rather, the subjective assessment of a teacher influenced by his relation to the pupil, the moment, or atmosphere, even by his or her current frame of mind. For some pupils this kind of assessment means their "small hell".

Today the function and necessity of systematic assessment and standardization of criteria and techniques is frequently discussed. Two functions of evaluation systems are stressed in the debate, the evaluative and the diagnostic. The proposals that have been forthcoming so far concern key stages of secondary schooling (at ages 15, 18).

Another problem is connected with the requirement of a school-leaving examination (maturita) for secondary schools, the social credibility of which has fallen as far as the established system is concerned.

Trends in curriculum policy

Debates on curriculum policy have led to the following two substantial conclusions:

(a) *Centralized and authoritative curriculum.* The previous experience of a strictly centralized and authoritative curriculum policy is quite unsuitable and lacks perspective. Its liberalization is necessary.

(b) *Loose curriculum frame at the central level.* It is possible to establish only a very loose curriculum frame at the central level (a list of subjects and the allocation of time for each as the common experience of many countries has shown). But this fact has been rejected by most schools and teachers. The main reasons for this reaction are a lack of experience and a state of unreadiness for the task of curriculum development by the schools and the teachers themselves.

So, the way forward in curriculum planning at the central level should be to develop:

a) a general common project involving the characteristics of education at different levels and for different types of schools, their values, functions and aims; the characteristics of their different environments, school organization, and participation in the school on the part of the parents and community.

b) a general core curriculum determining the number and the range of content for the compulsory subject areas, recommending optional and non-compulsory subjects and courses, and recommending policies for the organization of instruction (integration, block-instruction etc.)

c) a frame for the subject curriculum (brief syllabuses) involving the objectives, the core content, and recommendations concerning the structuring of teaching situations, learning projects, activities and examples.

Curriculum decisions at school level. This should be the next area of concern: Every individual school should make its own school curriculum for a 2 or 3-year period. It should take into account local conditions and possibilities. School curricula should acknowledge and support individual school priorities, action projects, and experiments and allow for the originality of the individual teacher.

The more detailed subject curriculum. This should fall within the competence and the responsibility of the teachers of the locality and the school.

Plans for future curriculum policy take into consideration the establishment of a system of educational standards and methods for their measurement. The incorporation of courses on curriculum development into teacher education is also currently being considered.

RESEARCH ON CURRICULUM IN CZECHOSLOVAKIA

First,I have to make an explanatory note: in the Czechoslovak tradition the complex concept of curriculum has not been accepted and until recently was even ignored.

Problems of curriculum research were split into the different aspects of research on instruction, which was traditionally conceptualized as "didactics". Consequently research on the curriculum as such was marginal. Problems of aims, goals, content and the means of education were investigated separately.

Since curriculum designers were mostly subject specialists, a comprehensive curriculum theory and a planned curriculum technology have not been developed. In spite of these facts some researchers continued to be concerned with analyses of curricular materials (syllabuses, textbooks) and with the real problems of the school curriculum (eg. J. Prucha 1988, E. Walterova 1990, 1991, a, b).

Now the new Institute of Educational and Psychological Research in the Faculty of Education, Charles University (established in 1990) has developed a research project called "Curriculum Innovations in Czech schools". This project is aimed at comparative research on the curriculum. This research deals particularly with problems and topics which are priorities in our educational reality, such as the diversification of the curriculum, curriculum policy, the content of education, textbooks, time as a curriculum category, the curriculum for gifted pupils, etc. Special interest is focused on problems regarding the European dimension in the curriculum.

In the future an institution specializing in curriculum research and development should be established. The idea is to delegate to this institution such activities as modelling the curriculum, helping schools to develop their own curriculum, and developing curriculum materials.

CURRICULUM DEVELOPMENT IN THE INTERNATIONAL CONTEXT

To describe and to compare current changes regarding the curriculum in the international context is extremely difficult. Even to identify the main tendencies is complicated, for many contradictions and discrepancies are apparent (OECD, 1990). Attempts to explain them accurately, and not over-simplify the real picture, has led to more detailed forms of description in which the general overview has to be accompanied by many examples, illustrations and details on specific issues (S. Rassekh, G. Vaideanu, 1987).

In spite of this, "very global tendencies crystalize" (W. Mitter, 1991) by which progress in curriculum development can be characterized (E. Walterova, 1991). These tendencies are:

- an accent on human values, the needs of personality development, contextual problems, the demands of current social life, the social environment and personal experience are the main influences on the curriculum.

- a more complex concept of curriculum and the priority of general aims (developmental, meta-cognitive and affective) as coordinators for learning situation design;

- a higher respect for differences between learners, leading to diversification within the curriculum, supportive programmes for groups and minorities, and individualized programmes;

- support for school autonomy, for alternative and original projects, and for cooperative projects;

- innovations in the curriculum: new subjects and courses, integrated courses and topics, (even attempts toward a strategy of "total curriculum area cover").

In countries where curriculum policy has been rather centralized the support for school autonomy in curriculum design has not been eliminated (for instance, Sweden, Finland, Italy, France). The central curriculum policy in some countries is aimed at the standardization of the curriculum and the development of national criteria for evaluation; examples are the national curriculum and the system of evaluation in England and Wales, the strategy of a national core curriculum in the USA, the development of common core curricula for secondary schools in the Netherlands.

CONCLUSION

With the changing situation in the world and the integrative processes taking place in Europe, particularly the development of the European community, common international problems emerge concerning curriculum development.

The dynamics of change lead to the need for the reconceptualization of the curriculum. There are, in addition, demands to take into account curriculum needs on the supranational level also.

Searching for the curriculum model that allows the curriculum to be controlled centrally but in a way that is participative encounters and involves designing a "humanistic core". Designing this "core" is also a task for comparative international research. Many initiatives and activities on different dimensions have been developed now in this area and the outlook is very hopeful. In fact, I believe that coordination and joint-venture research projects might support competent decision-making and bring about more effective results.

REFERENCES

Havel, V. (1990) : *O lidskou identitu* (Toward Human Identity). Praha, Rozmluvy .

Kotasek, J. et al. (1991) : *Soubor expertnich studii k rozvaze o skolstvi a vzdelanosti a jejich dalsim vyvoji v Ceskych zemich.* (Considerations about School and Education. Further Development in Czech Republic - Expert Studies). Pedagogická fakulta UK, Praha.

Mitter, W. (1991) : *School Reform in International Perspective: Trends and Problems. Pedagogika* 41, 1, pp. 7-23.

OECD: *Compulsory Schooling in a Changing World.* OECD, Paris 1983.

OECD: *Pathways for Learning.* OECD, Paris 1989.

Prucha, J. (1988): *Teorie a vyzkum skolnich ucebnic.* (Theory and Research on School Textbooks). SPN, Praha.

Rassekh, S. , Vaideanu, G. (1987) : *The Contents of Education. A worldwide view on their development from the present to the year 2000.* UNESCO, Paris.

Walterova, E. (1989) : *Aktualni trendy v projektovani vyuky.* (Current Trends in Curriculum Development). Pedagogika 39, 4, pp. 416-430.

Walterova, E. (1990) : *Projektovani stredni vseobecne vzdelavaci skoly.* (Curriculum Development in the Secondary Comprehensive School). USI, Praha.

Walterova, E. (1991) : *Problems of Transformation of Knowledge in Projects of Instruction.* In: Olkinuora, E.(ed.): *Knowledge Transmission Process in Finland and in Czechoslovakia.* University of Turku. Research Reports A: 148, a, pp. 20-38.

Walterova, E. (1991) : *Humanization of Education and Curriculum Development,* invited address delivered at 4th European Conference for Research on Learning and Instruction, Turku.

Walterova, E. (1991) : *Trendy v teori i tvorbe kurikula v zahranici.* (Trends in Curriculum Theory and Development Abroad). In: Prucha, J. (ed.): *Promeny vzdelani v mezinarodnim kontextu* (Transformation of Education in the International Context). Universita Karlova, Praha.

Expertni studie - strucne shrnuti. (Expert studies survey). Ucitelske noviny, 1992, 45, pp. 8-13.

Odpovedi ctyr tymu na 17 otazek o transformaci skolstvi. (Response of four teams to 17 questions concerning the school reform. Ucitelske noviny, 1992, 1, appendix pp. 1-4.

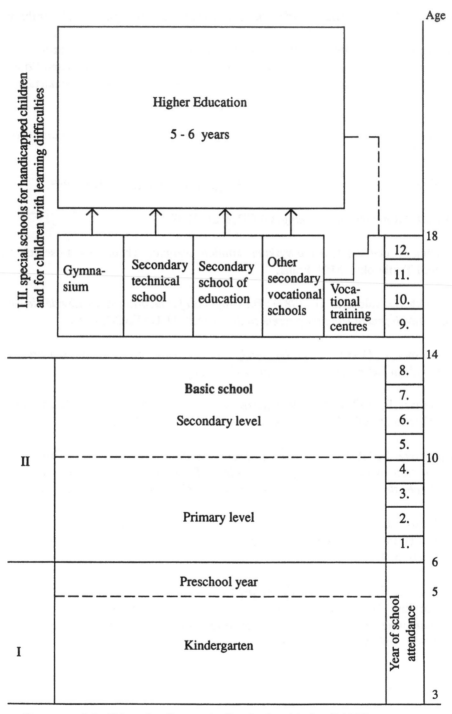

Structure of the Education System in Czechoslovakia (until 1990)

APPENDIX 2

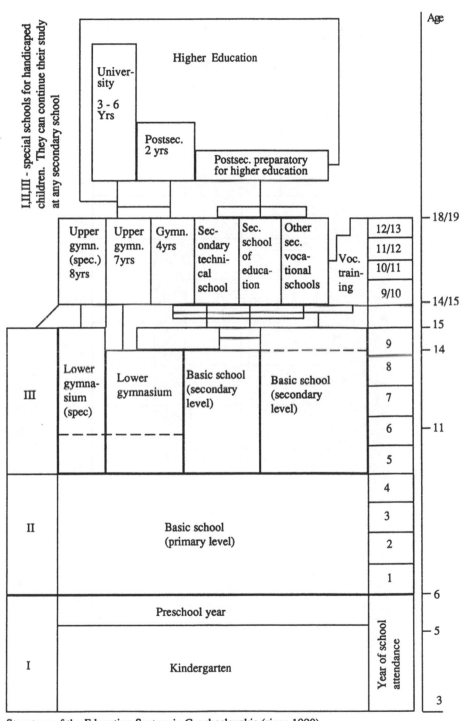

Structure of the Education System in Czechoslovakia (since 1990)

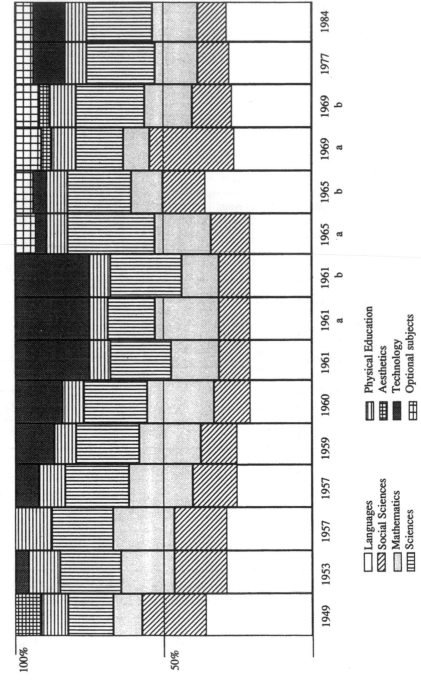

Czechoslovakia:
Curriculum of Secondary Gereral School (Upper level, age 14/15 - 17/18)
Development of time proportion

Curriculum: Table of weekly lessons

Ground school: secondary level (age 10 - 14) since 1990

Subject	Grade			
	5	6	7	8
Czech and literature	5-4	5-4	5-4	3-4
Foreign language	3-2	3-2	3-2	3-2
Mathematics	5-4	5-4	5-4	5-4
Civics	-	1	1	1
Physical education	3-2	3-2	3-2	3-2
Geography	2	x	x	x
History	2	x	x	x
Nature science	2	x	x	x
Music	1	x	x	x
Art	2	x	x	x
Physics	-	x	x	x
Chemistry	-	-	x	x
Crafts and technology	2		x	x
Optional subjects	-	-	x	x
Weekly	27-23	29-26	34-28	31-28
Non-compulsory subjects	2	2	2	2

x: Number of lessons is determined by school (principal)

Curriculum: Table of weekly lessons
Secondary general school (age 14-18) since 1990.

Subject	Grade			
	1	2	3	4
Czech and literature	3	3	3	3
Foreign language 1	3/2	3/2	3/2	3/2
Foreign language 2	2/3	2/3	2/3	2/3
Latin	x	x	x	x
Civics	x	x	2	2
History	2	2	2	x
Geography	2	2	x	x
Mathematics	3	3	2	2
Descriptive Geometry	x	x	x	x
Physics	2	2	2	x
Chemistry	2	2	2	x
Biology/Geology	2	2	2	0
Informatics	2	x	x	x
Aesthetics	2	x	x	x
Physical education	3	3	3	3
Optional subject 1	x	x	2	2
Optional subject 2	-	x	x	2
Optional subject 3	-	x	x	2
Optional subject 4	-	-	x	x
Weekly compulsory	28	24	25	21
Lessons decided by school	3	7	6	10
Together	31	31	31	31
Non-compulsory	x	x	x	x

x: Number of lessons is determined by school (principal).

RESEARCH INTO SECONDARY SCHOOL CURRICULA IN FINLAND

Erkki Kangasniemi and Päivi Tynjälä
Institute for Educational Research, University of Jyväskylä

This paper will briefly give some information and answers to the questions raised by the Council for Cultural Cooperation of the Council of Europe in its paper DECS/ Rech (91) 70 (Strasbourg, 8 October 1991).

RECENT REFORMS IN SECONDARY SCHOOL CURRICULA

In the 1980s the academic upper secondary school curriculum was organized into a modular system after experimental work in the 1970s. Contents in every subject are divided into courses. Each course consists typically of 38 lessons. Overall objectives have been formulated for each subject and specific objectives for each course.

In the modular system a student studies four to six subjects at a time. Subjects to be studied are changed four to five times during the school year.

Non-graded education is being tried out in the upper secondary schools. A non-graded curriculum removes grade level barriers. The aim is that individual students may experience continuous progress through the school programme and develop their abilities to the maximum. A student is allowed to make a choice if he/she wants to stay in school two to four, or even five years. A student has more freedom and flexibility to organize studies at his/her own pace than in the present system. Follow-up results show that students like the freedom and flexibility of the non-graded curriculum. However, there are many problems, especially in small schools in organizing 'nongradeness' in practice. Much more time is needed for study counselling. There is a cabinet decision, however, that all academic upper secondary schools will adopt the non-graded system in 1994. The current problems in the national economy may, however, delay this.

The time resource quota system (tuntikehys) is being carried out in secondary schools. The total amount of lessons that can be organized in each school depends on the amount of pupils they have. Within a national framework, the school decides how many lessons each subject will receive and how many classes there are in each subject in the school.

In the 1980s all post-compulsory vocational upper secondary education programmes were organized on 25 broad-basis orientations. Each of these begins with a one-year "basic course", after which the basic line is divided into about 10 "specialization

programmes" of one to four years (the number of all vocational programmes at the upper secondary level thus totalling around 250). The aims of the reform were:
- to broaden the scope of vocational curricula,
- to increase the amount of general/academic studies in vocational education and
- to improve the qualifications for further studies produced by the vocational programmes.

Flexible structure of vocational programmes.

Instead of the 250 centrally planned vocational programmes/curricula (see above), more flexible ways of composing vocational studies are being sought.

The aim is:
- to strengthen the role of vocational schools and institutes in adjusting their vocational curricula to the needs of the students and of the changing world of work/ production
- to allow students more freedom in composing their study programmes and in timing the vocational as well as academic courses included in their study programmes,
- to develop new ways of organizing and modulizing the vocational studies into meaningful study units, and
- to develop new ways of integrating maths and science studies with vocational studies.

Upper secondary education reform ("youth education" experiment 1992-1999) was initiated by creating new forms and possibilities for local cooperation between vocational schools and academic upper secondary schools. An attempt is being made:
- to lessen the strict differentiation between these two types of upper secondary education,
- to increase the share of common subjects in academic and vocational programmes (mother tongue, mathematics, two foreign languages plus alternative courses in science and social studies for all),
- to promote integration and mutual enrichment between vocational and academic studies, and
- to offer the students new possibilities to choose their personal combination of academic and vocational studies.

SELECTION OF CURRICULUM CONTENT

Curriculum construction has not been systematically studied in Finland. Therefore, it is not possible to give any definite answers to the questions posed in the Conference memorandum. However, it can be stated that when the comprehensive school curriculum of 1970 was prepared there was quite extensive discussion about the "determinants" of the curriculum. These were considered to be (i) society and social requirements, (ii) subjects and central topics in each subject, and (iii) the pupils. A harmonious balance between the three determinants was considered important.

It is generally agreed that, in practice, subjects and issues regarding subjects dominated the process of curriculum construction. Similarly, it was breadth that was emphasized. Early studies in lower and upper secondary (academic) schools showed that there was a considerable variance in pupil achievement and that only part of the taught subject-matter was learned.

This led to work on proposals for core curricula in the mother tongue, foreign languages and mathematics in the lower secondary school (upper level of the comprehensive school). These proposals were circulated to all schools but they were not officially ratified. This work did have some impact, however, on subsequent curriculum construction in the early 1980s.

The academic upper secondary school curriculum was organized into courses as mentioned above. Achievement is assessed after each course. One of the advantages of the course-based curriculum was considered to be concentration on a fewer number of subjects at any particular time of the school year.

In vocational education, the common core is limited to the general subjects of the first year (basic course) where courses in the mother tongue, mathematics, one foreign language, and physical education cover some 30% of the studies. Vocational breadth is emphasized within each of the basic orientations (by having a common first-year course within the branch).

ORGANIZATION OF THE CURRICULUM

The lower secondary and upper secondary (academic) school curriculum is basically organized in terms of subjects. There is a national timetable which establishes the number of lessons for each subject and the permissible local variation. There has been a move to integrate subject matter across subjects. This is usually based on themes. There is also increasing effort to forge closer links between the school and the external community. For instance, all pupils at the lower secondary level spend a week at some workplace of their choice.

The current plan is that in 1994 there will be new national curricula for lower and upper secondary (academic) schools. Then schools may also be able to adapt their own curriculum accordingly. The policy is that the curriculum will be (or at least it is planned to be) finally a school-based curriculum. There is a movement towards abolishing the strong national control of the past. Differences in content and outcomes of education are thus likely to increase.

In vocational education most of the curriculum is organized in vocational subjects and workshop practice (mostly taking place at schools, very little in firms and factories). New forms of organizing the vocational curricula are being sought by the flexible-structure experimentation programme. The study programmes are planned to be composed of general subject courses, branch-specific basic courses, major vocational courses and subsidiary vocational courses.

CERTAIN TOPICS IN CURRICULA

In lower secondary school and upper (academic) secondary school most of the subjects are obligatory for all pupils and the share of optional subjects is relatively small. There is, however, a proposal by a Minister of Education working party according to which flexibility and students' freedom to choose subjects and courses would increase substantially. The decision on new school schedules will be made in autumn 1992.

Nowadays secondary school curricula are also quite extensive. Religion is a common subject and it covers all large world religions and includes ethical and moral questions. Those who are not willing to study religion may choose "life and ethics" (elämänkatsomustieto) which includes ethical and moral questions.

Health education, drug education, sex education and similar topics are dealt with in physical education. Intercultural education is one of the objectives in the teaching of all foreign languages. There is no specific subject for education for democratic values or gender equality but themes related to them are discussed through the curriculum e.g. in history, civics and mother tongue. Computer studies are a popular elective subject and computers are increasingly integrated in normal subject studies.

For students interested in music, arts or sports there are specialized orientations and schools on the upper secondary level.

IMPACT OF CHANGES IN THE ECONOMY AND WORK LIFE

By broadening the content of vocational education programmes, the reform of the 80s aimed at an increase in the capacity of the trained work force to cope with the changing qualification requirements of working life. This aim is pursued further by the flexible-structure experimentation programme as well as by the youth education experimentation programme. Both of these aim at an increase in the local and individual flexibility of the vocational education programmes, and at an improvement of the base for further/continuing studies of the labour force.

Separate courses in information technologies were introduced in the 80s in some vocational education programmes. The current trend is towards integrating the new technologies into the relevant vocational courses.

LONGITUDINAL CAREER RESEARCH

Research on the career development of vocationally educated youth in the 80s has shown a rather low correspondence between the branch of training and the branch of employment after transition to working life.

Ongoing research evaluates and compares the post-secondary work and education careers of students with vocational vs. general (academic) upper secondary education.

APPENDIX - RESEARCH PROJECTS AND PUBLICATIONS BY PROJECT

Curriculum Development for Primary and Secondary Education
Paavo Malinen
University of Jyväskylä, Department of Teacher Education
PO Box 35, SF 40351 Jyväskylä

Publications:
Malinen, P. 1987. Perspectives on the Operation of School Curricula. Scandinavian Journal of Educational Research 31, 71-80.
Malinen, P. 1987. The Operational Structure of School Curriculum. In P. Malinen & P. Kansanen (eds.): Research Frames of the Finnish Curriculum. University of Helsinki, Department of Teacher Education. Research Report 63, 69-94.
Malinen, P. 1989. How Can 'Lehrplan' and 'Curriculum' Be Combined? In A. Korner (Hrzg.): Bildungspolitische Perspektiven in Nord- und Westeuropa. Studien zum Bildunsgwesen Nord- und Westeuropas, Band 9. Verlag der Ferberschen Universitätsbuchhandlung, 69-77.
Malinen, P. 1985. Curricula in Current Education. (Opetussuunnitelmat nykyajan koulutuksessa). Helsinki: Otava. (in Finnish)

Research on the Development of Upper Secondary School Teaching
Jouni Valijarvi
Institute for Educational Research, University of Jyväskylä
PO Box 35, SF-40351 Jyväskylä

Publications:
Välijärvi, J. 1980. The Structure of the New Upper Secondary School Curriculum as Rated by Teachers and Pupils. University of Jyväskylä. Institute for Educational Research. Bulletin 150. (in Finnish, English abstract)
Välijärvi, J. 1988. Tradition and Reform in Liberal Education. (Jaksomuotoinen opetus lukion ylissivistävän tradition kehittäjänä.) University of Jyväskylä. Institute for Educational Research. Publication Series A. Research Reports 15. (in Finnish, English abstract).
Välijärvi, J. 1990. Changing General Education in Finland. Life and Education in

Finland 1 (2), 6-9.

Huttunen, I. & Välijärvi, J. 1990. Reforming the learning environment of the upper secondary school by making use of non-graded teaching. (Lukion oppimisympäristön uundistaminen luokattumuutta hyödyntäen.) Helsinki: The National Board of Education. (in Finnish).

Publications on research into the time resource quota system in lower secondary and upper secondary school:

Suortti, J., Nikkanen, P. & Jokinen, H. 1983. The Time Resource Quota System as a Means of Developing the Upper Level of the Comprehensive School. University of Jyväskylä. Reports from the Institute for Educational Research 328. (in Finnish, English summary).

Nikkanen, P. 1986. Follow-up of Streamed and Non-Streamed Teaching in the Finnish Comprehensive School. University of Jyvaskyla. Reports from the Institute for Educational Research 369. (in Finnish, English summary).

Nikkanen, P. & Mehtäläinen, J. 1983. Tutor-method. University of Jyvaskyla. Reports from the Institute for Educational Research 334. (in Finnish, English summary).

The Ministry of Education: The Planning Group of The Time Resource Quota System. 1985. The Time Research Quota System in Comprehensive School. (Tuntikehysjärjestelmä peruskoulussa.) (in Finnish).

Publications on the realization of the curriculum:
Kansanen, P., Uusikylä, K. & Kalla, H. 1980. Realization of the Curriculum: the Starting Points of the Research. (Opetussuunnitelman toteutuminen: tutkimusprojektin lähtökohdat). The National Board of Education. Research Reports 35. (in Finnish).

Mother Tongue Curricula in Different Countries
IEA Study of Writing Curricula
Hannu Saari
Institute for Educational Research, University of Jyväskylä
PO Box 35, SF-40351, Jyväskylä

Publications
Saari, H. 1991. Writing Curricula in Sixteen Countries. International Study in Written Composition. University of Jyväskylä. Institute for Educational Research. Publication series A. Research reports 42.

International Reading Literacy Study (IEA)
Pirjo Linnakylä
Institute for Educational Research, University of Jyväskylä
PO Box 35, SF-40351, Jyväskylä

A Follow-up Study on Mother Tongue Teaching

Katri Sarmavouri
University of Turku, Department of Teacher Education
Lemminkäisenkatu 1, 20520 Turku

Publications:
Sarmavuori, K. 1986. Questioning Methods in Teaching of the Mother Tongue (Finnish). Scandinavian Journal of Educational Research 30, 141-151.

Publications on mother tongue curricula in vocational schools
Hirvi, V. 1982. First-year Mother Tongue Curriculum in Secondary-Level Vocational Schools and its Feasibility as Rated by Teachers of General Vocational Schools. University of Jyväskylä. Reports from the Institute for Educational Research 314. Dissertation. (in Finnish, English Summary).

Publications on foreign language curricula:
Takala, S. 1980. New Orientations in Foreign Language Syllabus Construction and Language Planning: a Case Study of Finland. University of Jyväskylä. Institute for Educational Research. Bulletin 155.

* The Second International Study of Achievements in Mathematics (IEA/SIMS)
Erkki Kangasniemi
Institute for Educational Research, University of Jyväskylä
PO Box 35, SF 40351 Jyväskylä

Publications:
Kangasniemi, E. 1989. Curriculum and Student Achievement in Mathematics. University of Jyväskylä. Institute for Educational Research. Publication Series A. Research reports 28. (in Finnish, English summary)

* Research on Mathematics Education/The Learning and Teaching of Concepts in School Mathematics
* The National Assessment of the Finnish Comprehensive School 1990/Mathematics
Pekka Kupari
Institute for Educational Research, University of Jyväskylä
PO Box 35, SF-40351 Jyväskylä

Publications:
Kupari, P. 1989. Applications in Finnish School Mathematics Education - Research Results and Development Prospects. In W. Blum, J.S. Berry, R. Biehler, I.D. Huntley, G. Kaiser-Messmer & L. Profke (eds.) Applications and Modelling in Learning and Teaching Mathematics. Chichester: Ellis Horwood, 88-91.
Kupari, P. 1989. Problem Solving, Modelling and Applications in the Finnish School Mathematics Curriculum 1970-1985. In W. Blum, M. Niss & I. Huntley (eds.) Modelling, Applications and Applied Problem Solving: Teaching Mathematics in a Real Context. Chichester: Ellis Horwood, 233-241.

Kupari, P. 1989. Some Features in the Development of Geometry Teaching. In E. Pehkonen (ed.) Geometry Teaching - Geometrienunterricht. Conference on the Teaching of Geometry in Helsinki 1. 4.8. 1989. University of Helsinki. Department of Teacher Education. Research Report 74, 171-181.

Development of the Teaching of Mathematical and Science Subjects in Vocational Education

Virpi Koponen
University of Jyväskylä, Institute for Educational Research
PO Box 35, SF-40351, Jyväskylä.

The Second International Study of Achievements in Science Education (IEA/SISS)

Kimmo Leimu
Institute for Educational Research, University of Jyväskylä
PO Box 35, SF 40351 Jyväskylä

Achievements of Science Education in Finland
Science Education Research in Finland/The National Assessment of the Comprehensive School in Finland 1990

Jari Lauren
Institute for Educational Research, University of Jyväskylä
PO Box 35, SF-40351 Jyväskylä

Publications:

Lauren, J. 1989. Science Education in Finland. Yearbook 1987-88. University of Jyväskylä. Institute for Educational Research. Publication Series B. Theory into Practice 36.

Publications on physics curricula in upper secondary school:

Erätuuli, M. 1983. The Development of the Curriculum of Upper Secondary School Physics in Finland 1916-1979. (Lukfiofysiikan opetussuunnitelman kehittyminen Suomessa vuosina 1916-1979). University of Helsinki. Department of Education. Research Reports 83. Dissertation. (in Finnish)

Pedagogic Development in Vocational Institutions of Education

Jorma Ekola
Institute for Educational Research, University of Jyväskylä
PO Box 35, SF-40351 Jyväskylä

Publications:

Ekola, J. (Ed.) 1991. Reform of Vocational Secondary-Level Education. Evaluations of the Implementation of the Reform. University of Jyväskylä. Institute for Educational Research. Publication Series B. Theory into Practice 56. (in Finnish,

English abstract).

Ekola, J., Vuorinen, P. & Kämäräinen, P. 1991. The Vocational Education Reform in the 80s. Ammattikasvatushallituksen tutkimuksia ja selosteita no: 30/1991. Helsinki: Valtion painatuskeskus. (in Finnish).

Mutual Enrichment of Academic and Vocational Education at the Upper Secondary Level
Matti Vesa Volanen
Institute for Educational Research, University of Jyväskylä
PO Box 35, SF-40351 Jyväskylä

Monitoring of the Experimentation of Flexible Schooling Structures in the Field of Social Services
Heli Nurmi
Institute for Educational Research, University of Jyväskylä
PO Box 35, SF-40351 Jyväskylä

Project Method in Secondary School Teaching
Pekka Penttinen
University of Jyväskylä, Department of Education
PO Box 35, SF-40351 Jyväskylä

Education and Gender
Elina Lahelma
University of Helsinki, Center of Further Educational in Vantaa
Asematie 11B, SF-01300 Vantaa

Publications:
Lahelma E. & Ruotonen, H. 1986. Equality of the Sexes in School and Teacher Education in Finland. Paper presented in the 11th Conference of The Association for Teacher Education in Europe. Workshop Teacher Education for Equal Opportunities for girls and boys. Toulouse 1.9.1986.

Lahelma, E. 1988. Teachers and Projects of Equal Opportunities in Schools: Reflections from the Experience in Finland. Paper presented in the 13th Conference of The Association for Teacher Education in Europe. Workshop on Teacher Education for Equal Opportunities for Girls and Boys. Barcelona 5.9.1988.

Lahelma, E. 1990. Curriculum and Gender Agenda: Some Examples from Finland. Paper presented in the 15th Conference of The Association for Teacher Education in Europe. Workshop on Teacher Education for Equal Opportunities for Girls and Boys. Limerick 30.8.1990.

Publications on curricula for religious instruction:
Kalevi Tamminen et al. 1982. Research on Religious Pedagogy and the Development of the Curriculum of Religion (Uskonnonpedagoginen tutkimus ja uskonnon opetussuunnitelman kehittäminen: tutkijatyöryhmän arviointi

peruskoulun evankelisluterilaisen uskonnon opetussuunnitelmasta ja näkökohtia sen kehittämiseksi.) University of Helsinki. Institute of Practical Theology. Publications of Religious Pedagogy A 19. (in Finnish).

Pyysiäinen, M. 1982. Confessional, Non-confessional and Objective Religious Instruction: a Curriculum Analysis of Religious Instruction in the Comprehensive Schools of Finland and Sweden. Helsinki: Kirjapaja. Dissertation. (in Finnish, English abstract)

Publications on mass-media education:

Minkkinen, S. 1980. Curricula for mass - media - education. (Joukkotiedotuskasvatuksen opetussuunnitelmat) University of Tampere. Department of Communications Theory and Mass Media. Teaching material no 34. (in Finnish)

Asikainen, K. et al. 1982. Mass-media Education in Lower Secondary School. (Peruskoulun yläasteen viestintäkasvatus) Jyväskylän kaupunki: Kasvatusprojekti II. (in Finnish)

Asikainen, K. et al. 1982. Mass-media Education in Upper Secondary School. (Lukion viestintäkasvatus). Jyväskylän kaupunki: Kasvatusprojekti II. (in Finnish).

Publications on international education:

The National Board of Vocational Education. 1982. 1984. A Plan for the Implementation of International Education in Vocational Schools. (Suunnitelma kansainvälisyyskasvatuksen toteuttamiseksi ammatillisissa oppilaitoksissa./ Ammattikasvatushallituksen kansainvälisyyskasvatustyöryhmän mietintö). (in Finnish)

Hietala, M. 1982. World View in History Textbooks: the Realization of Unesco's Recommendation for International Education in History Textbooks Used in the Finnish Upper Secondary School. (Maailmankuva historian oppikirjoissa: Unescon kansainvälisyyskasvatussuosituksen toteutuminen soumalaisissa lukion historian oppikirjoissa.) University of Helsinki. Department of History. Publication no 9.

Publications on health education:

Rostila, I. 1983. The Teacher's Point of View. Health Education at Vocational Schools (Yleisten ammattikoulujen terveyskasvatus terveystiedeon opettajien näkökulmasta.) Helsinki: Lääkintöhallituksen julkaisuja, Terveyskasvatus. (in Finnish, English abstract).

Korhonen, J. & Myllymäki, J. 1988. Health Education in the Course-based Upper Secondary School in School Year 1984-85. (Terveystiedon opetus kurssimuotoisessa lukiossa lukuvuonna 1984-85). University of Jyväskylä. Department of Health Sciences. Series A. Research reports 4. (in Finnish)

Computer Studies and Computer-Assisted Teaching in Comprehensive School

and in Upper Secondary School
Raimo Konttinen (et al.)

Raimo Konttinen, University of Jyväskylä, Institute for Educational Research
PO Box 35, SF-40351 Jyväskylä

Publications on computer studies:
The National Board of General Education, 1988. Computer-assisted Education in Upper Secondary School. (Lukion tietokoneavusteinen opetus: kokeilun loppuraporttien yhteenveto) Kouluhallitus: kokeilu- ja tutkimustoimisto. (in Finnish)

Lieno, J. 1990. Computers in the Development of Teaching. Project Studies in the Lower and the Upper Secondary School. (Tietokone opetuksen kehittämisessä. Projektiopiskelu yläasteella ja lukiossa.) University of Helsinki. Department of Education. Research reports 130. (in Finnish)

The Ministry of Education. 1989. The Integration of Information Technology into School Teaching: Evaluation and Further Measures. (Tietotekniikan integroiminen kouluopetukseen: tulosten arviointi ja jatkotoimet. Opetusministeriö. Tietokone opetuksessa - johtoryhmä.) (in Finnish)

Physical Education and Nature - The Pedagogical Research Area
Martti Silvennoinen & Risto Telama

University of Jyväskylä, Department of Physical Education
PO Box 35, SF-40351, Jyväskylä

Publications:
Silvennoinen, M. 1990. Outdoor Physical Education - New Aspects. Conference Proceedings, Vol 1. Auckland: NZAHPER, 195-202.

Publications on art education:
Tuominen, E. 1990. Art Education in Vocational Schools in Finland (Taidekasvatuksen opetus maamme ammattioppilaitoksissa). Vocational Teacher Training Institute in Hämeenlinna. (Hämeenlinnan Ammatillinen Opettajakorkeakoulu) Publications 59. (in Finnish)

Curricular Strategies for Lifelong Learning
Eero Rope, Tero Autio, Raimo Jaakkola & Pekka Kämäräinen

University of Tampere, Department of Education
PO Box 605, SF-33101 Tampere

Publications:
Ropo, E. 1990. Cognitive Learning as a Basis for New Curricular Strategies. Paper presented in the Conference of Educational Research in Finland 22.-24.11.1990.

Autio, T. 1990. Theory of Lifelong Education from the Viewpoint of Educational Philosophy. Paper presented in the Conference of Educational Research in Finland 22.-24.11.1990.

Jaakkola, R. 1990. Dimensions of Curricular Strategies for Lifelong Learning: Preliminary Questions. Paper presented in the Conference of Educational Research in Finland 22.-24.11.1990.

Kämäräinen, P. 1990. On the Issues of Structural Continuity and Curriculum Development within Current Educational Reforms - Some Introductory Remarks. Paper presented in the Conference of Educational Research in Finland 22.-24.11.1990.

RESEARCH INTO SECONDARY SCHOOL CURRICULA IN IRELAND

Sandra Ryan and Thomas Kellaghan
Educational Research Centre, St. Patrick's College, Dublin

Secondary schools in Ireland cater for students between the ages of 12 and 18 years. While education is not compulsory for students over 15 years of age, 90% of 16-year-olds and 74% of 17-year-olds attend school (Ireland: Department of Education, 1991). Students are distributed among three main types of school: traditional academic secondary schools (62.4%), vocational schools (24.9%), and comprehensive/community schools (11.7%) (A further 1% attend other types of school). Up to the 1960s, curricula in academic secondary schools were more academic than curricula in vocational schools. With the introduction of comprehensive schools in the 1960s, the distinction between schools in what curricula they could offer was abolished. Today all types of schools are encouraged to offer comprehensive curricula. In practice, however, many schools continue to exhibit their traditional orientations.

The comprehensivization of curricula in the 1960s served to increase the emphasis on technical and 'vocational' subjects in school curricula. It was felt that this would have economic benefits (preparing a skilled work force) as well as personal ones (catering for a wide range of student abilities and interests).

Curricula are organized on the basis of subjects or disciplines. Schools can choose from about 30 subjects. The majority of students study eight or nine subjects during their first three years in a second-level school and seven or eight during their final two or three years. The most popular subjects are English, Irish, Mathematics, French, Science, History, and Geography. It has been estimated that during the period of compulsory full-time education (which includes primary schooling as well as lower secondary grades) Ireland is above the average for countries in the European Community in the percentage of time which is allocated to language, about average in the time that is allocated to scientific subjects, social sciences, technologies, and ethics and religious instruction, and below average in allocation of time to artistic subjects and physical education (EURYDICE, 1987).

GENERAL CURRICULUM

There is relatively little research which has addressed the question of type or types of curricula that seem most appropriate for students in second-level schools at a time when participation rates are increasing dramatically in such schools and the conditions of work outside school are subject to rapid change (Mulcahy, 1981, 1989). In practice, in response to government policy, there has been an increase over the past two decades in the proportion of students who are enrolled in technical vocational courses (Lewis & Kellaghan, 1987) and it seems to be generally assumed that more students should be encouraged to follow the technical vocational route in future (Ireland, 1992; Roche & Tansey, 1992). If the form of education which is presented as technical vocational in schools in the future is similar to the form that is presented at the moment, then it will be a type of education that is designed not to provide students with specific vocational skills but with general competencies and values which can later be applied and developed in a wide range of occupations and other adult situations (Kellaghan & Lewis, 1991).

An approach in which a broad-based vocational education is offered in schools is not inconsistent with industrialists' perceptions of the characteristics which school leavers should possess. In a survey of 150 member companies carried out by the Confederation of Irish Industry (1990), in which respondents rated the importance of school leavers' skills for employment, 70% rated oral communication skills as 'very important'; a similar rating was given by 58% to written communication, by 48% to numeracy, by 46% to enterprise/initiative, and by 39% to problem-solving.

In another survey, young people who had left school five years previously also almost universally regarded basic education (in literacy and numeracy) as important for working life. However, they also saw a role for technical vocational subjects. This was particularly true for less qualified individuals (Hannan & Shortall, 1991).

SPECIFIC CURRICULA

Research in individual subject areas has been carried out on mathematics in the context of international studies. In these studies, curricula have been analysed to highlight cross-national similarities and differences (Travers & Westbury, 1989). The analysis revealed that the educational system in Ireland (together with a few other countries) had been heavily influenced by the 'new mathematics' of the 1960s based on the Bourbaki tradition, which placed a heavy emphasis on algebra and modern symbolism. Comparison with curricula in other countries indicated that Irish acceptance of the 'new mathematics' had been more enthusiastic than elsewhere with the result that the mathematics curriculum in Irish schools was out of line with the curricula of most other countries. Further, it appeared that the mathematics curriculum in Irish schools attempted to cover too much at too early a stage. These defects have been addressed in recent reforms.

EDUCATION AND LATER EMPLOYMENT

As in other countries, educational level (the highest school grade completed) has been found to relate to later occupational status in Ireland. The longer one stays in full-time education, the higher the status of the occupation one is likely to enter (Greaney & Kellaghan, 1984). Not surprisingly, former students' satisfaction with their education is also related to the students' level of educational attainment

(Hannan & Shortall, 1991).

Not only is the number of years completed in school important for one's future occupational status but employers seem to take into account performance on public examinations (taken at about 15 and 17 years of age) in making employment decisions. Students who do not take any public examinations are much more likely to be unemployed than students who are more successful in the educational system (Ireland: Department of Labour, 1991). So also are students who perform poorly on such examinations (Hannan, 1992). Specialization in technical vocational subjects only increases slightly the probability of employment in one's own locality (Hannan, 1992).

Young people are not unaware of these relationships. A study of individuals who had left school five years earlier revealed that they thought that their level of educational certification was an important factor in getting a job. Those without a job attributed a greater importance to qualifications in gaining employment than did those with a job (Hannan & Shortall, 1991).

REFERENCES

Confederation of Irish Industry (1990). *Human resources - The key issue.* Confederation of Irish Industry Newsletter, 53(9), 1-7.

EURYDICE (1987). *Basic education and competence in the member states of the European community,* Brussels. Author.

Greaney, V., & Kellaghan, T. (1984). *Equality of opportunity in Irish schools.* Dublin: Educational Company.

Hannan, D.F. (1992), *Education, employment and local economic development,* Paper read at Regional Studies Association Conference, Dun Laoghaire, Ireland, March 28.

Hannan, D.F. & Shortall, S. (1991). *The quality of their education.* Paper No. 153. Dublin: Economics and Social Research Institute.

Ireland (1992). *Education for a changing world. Green paper on education.* Dublin: Government of Ireland.

Ireland: Department of Education (1991). *Statistical report 1989/90.* Dublin: Stationary Office.

Ireland: Department of Labour (1991). *Economic status of school leavers.* Dublin: Author.

Kellaghan, T., & Lewis, M. (1991). *Transition education in Irish schools.* Dublin: Educational Company.

Lewis, M., & Kallaghan, T., (1987) *Vocationalism in Irish second-level education, Irish Journal of Education,* 21, 5-35.

Mulcahy, D.G. (1981). *Curriculum and policy in Irish post-primary education.* Dublin: Institute of Public Administration.

Mulcahy, D.G. (1989). *Official perceptions of curriculum in Irish second-level education,.* In D.G. Malcahy and D. O'Sullivan (Eds.), *Irish educational policy.*

Policy and substance. Dublin: Institute of Public Administration.

Roche, F., & Tansey, P. (1992). *Industrial training in Ireland*. Dublin: Stationery Office.

Travers, K.J., & Estbury, I. (1989). *The IEA study of mathematics 1: Analysis of mathematics curricula*. Oxford: Pergamon.

SECONDARY EDUCATION REFORM IN LUXEMBOURG

Jeannot Hansen
Head of the Luxembourg Delegation to the Education Committee
Ministère de l'Education Nationale

Since 1968 when the last major reform took place in secondary education in Luxembourg, there have been considerable changes in the economic, technical and social sectors as well as in the universities, which make it essential to adapt the school to the requirements of the modern world.

It is not a question of sweeping away the whole education system but rather of retaining the best of it while adjusting to partly new imperatives. The fundamental precepts of Luxembourg's secondary education are therefore not questioned and its purpose remains essentially to prepare students for advanced studies on the basis of a broader cultural framework. Nevertheless, the education provided must also enable young people not pursuing higher education to find a job. They will be better equipped for this if the school has trained them to cope with the requirements of a fluctuating labour market.

The main lines of the reform envisaged are the following:

a. Curricula must make pupils capable of understanding the contemporary world in its complexity and diversity.

In view of the changes we have seen and those confronting us in the future, we must encourage versatility in the teaching provided, give general education greater importance by prolonging it as far as possible and also by delaying the moment of specialization. Again, we must redefine what we call "general education" and "specialization". Thus familiarity with the new technologies must form part of the stock of knowledge of all secondary school pupils.

b. Entry of pupils into either the university or the world of work, both of which are constantly evolving, must be made easier.

Improvements in the quality of teaching, a major asset in confronting the challenge posed by the opening up of the internal market in 1992, must be effected by increased diversifications of provision, in keeping with the requirements of the university sector and the demands of the labour market. Optional subjects will be offered in all secondary schools ("lycées") in addition to the basic subjects and will provide valuable supplementary training, on the understanding that the weekly timetable must not be overloaded. Young people will be more adaptable in a changing situation.

If the weekly timetable for basic and specialist lessons amounts to fewer than 30 hours, pupils will have to choose some supplementary subjects. So they will be more closely associated than in the past in designing their own school and career prospects. This freedom should increase pupils' motivation and the quality of their education.

c. Pupils must learn to improve their verbal communication, to take the initiative and to work in a team. Teaching methods must bear in mind this need and attach greater importance to oral presentations, personal research and group activities.

d. It is desirable for close and systematic contacts to be established between secondary education and the universities and for there to be real cooperation with the world of work and business. Information on change in either of these domains must be conveyed first through the teaching profession. This means that in-service training for teachers - according to methods to be determined - will be even more necessary if we wish to give our pupils the best possible chance. It would seem useful and even essential to organize visits and placements in firms and to find ways of making company managers aware of the real situation confronting teachers in schools.

This new approach to studying must allow the pupils to find the right path for themselves and at the same time to acquire a strictly logical approach and an open and flexible attitude.

The faculty to analyse and synthesise will be developed along with the capacity to express - not only correctly but clearly and precisely - the thoughts which stem from a culture which is all the wider for being rooted in the plurilingualism of our country.

To give better guidance for pupils, the present system of clearly specialized sections beginning with class IV will be replaced by a system reducing early and virtually irreversible selection of options. Secondary education will consist of two divisions:
- the lower division comprising classes VII, VI and V;
- the upper division, divided into two sections:
  a general section (classes IV and III);
  a specialist section (classes II and I).

When the pupils begin the general section (IV, III) they must choose their general stream and opt for either the arts or the science stream which differ essentially in having different mathematics courses. Care will be taken to improve the oral competence of all pupils, along with their written competence. Timetable and curriculum differences between so-called classical and modern education will be less marked than in the past.

Pupils moving into class IV choose their general stream, arts or science, but do not yet have the opportunity to influence considerably their course composition. From this class onwards, however, there will be pre-specialization options: languages, art, music, mathematics, economics and sciences.

The differences between the teaching given in the arts and the science streams are not very great and pupils who have chosen the wrong stream will be able to change much more readily than at present: to change they will only have to pass one or two extra tests in the pre-specialization subjects. Changes within the stream will also be easier.

On entering the specialization section (classes II and I), pupils will choose between seven options with either an arts or a scientific bias which will take more broadly into account each pupil's particular talents and the requirements of the contemporary world. Pupils will opt to study languages, the humanities and social sciences, the arts, mathematics, sciences or economics more thoroughly. Less time will be spent on common courses and specialization will be reflected in a range of options slightly different from those available at present.

Arts stream:
   A1 languages
   A2 humanities and social sciences
   E art
   F music

Sciences stream:
   B mathematics - physical sciences
   C mathematics - natural sciences
   D mathematics - economic sciences

In the specialization years, it will still theoretically be possible to change streams but this will be harder because of the constantly increasing level of specialization.

The number of compulsory lessons weekly will vary according to the stream and option chosen by pupils. If possible, it will be kept below 30 lessons. Most pupils will therefore have to choose at least one supplementary course.

The options system, which will allow an individual school profile to be chosen, will include the following variations:
- pre-specialization options in classes IV and III;
- supplementary options in classes II and I;
- free options which, as in the past, will be offered to pupils wishing to improve their knowledge or to use their spare time constructively in addition to the strictly curricular activities.

Since supplementary optional subjects are an important innovation in our educational system, I shall describe this at greater length. These are courses in which one or two lessons a week are spent studying subjects related to the general or specialized education of the pupil and are additional to the subjects studied otherwise. The class council, the educational psychology and counselling service, in consultation with the school principal, will assist pupils, after studying their school record card, to choose options in keeping with their ability and likely to give them the best chance for continuing their studies or choosing a career.

The extra optional subjects on the curriculum will be suggested by national commissions and will be essentially the same for all lycées. Where there is a specific demand, however, particular courses may be instituted in particular schools, taking advantage of the special expertise of teachers or existing infrastructures. The intrinsic value of the subject and the interests of the pupils will form the basis for setting up extra optional courses.

In the wake of the reforms envisaged, the criteria for promotion from one class to the next must be reviewed and redefined. It is obvious that the importance of the promotion index attributed to the subjects taught in the 30 or 31 lessons per week will vary considerably according to whether or not the subject is essential for the pupils' selected stream.

These changes in the structure of secondary education, in curricula and in assessment of pupils' efforts will be accompanied by changes in teaching methods which must take into account as far as possible new requirements formulated in educational circles and in the world of work. Pupils emerging from secondary schools who will later occupy posts of responsibility must be capable, for example, of communicating verbally, taking the initiative and working in a team. Without relinquishing the advantages offered by traditional teaching, directed chiefly at the individual, these talents, too often left to languish, must also be developed.

Judicious and more systematic use must be made of audio-visual resources and the new information technologies available to us through advanced technology: teaching will be livelier and pupils will take a more active part.

This will all require the assistance of the teaching profession. In-service training for teachers is absolutely essential for implementing the reform.

Timetabling and syllabuses for each subject, to be fixed by ministerial decision, raise difficult problems. Luxembourg must prepare secondary school pupils to continue with their studies in both German-speaking and French-speaking areas, in addition to catering for those who register with English, American, Spanish, Italian or other universities. Thorough knowledge of several languages is both an opportunity and a burden which we have to accept. Since it is necessary to keep the weekly lesson load under 30 or 31, it is inevitable that there have to be sacrifices in one subject or another, on either the arts or the science side.

The world around us is changing fast and the school should prepare pupils to face the changes while not abandoning the best of its past achievements.

RESEARCH INTO SECONDARY SCHOOL CURRICULA IN THE SLOVAK REPUBLIC

Vladimir Burjan
Head of the Group for the Evaluation of Achievement
Institute for Education Research, Bratislava, Slovakia

BASIC BACKGROUND INFORMATION

The state-of-the-art of research about the secondary school curricula in Slovakia[1] can hardly be understood without keeping in mind the whole context of the transition period our country is now undergoing.

The impact of the fundamental political changes initiated in Czechoslovakia in November 1989 on educational research in general was outlined in our paper prepared for a similar CDCC workshop in San Marino. That paper contains also an extensive list of those research problems which are currently being considered as most relevant and which are being given most attention and which are most often stressed in educational research in Slovakia. Let us here just briefly summarize the main characteristics of the present situation.

(a) Our most important, and at the same time most difficult, present task is to overcome the explicit as well as the implicit heritage of the "communist educational paradigm" which was characterized by
- a strict control of the communist party over the whole educational system resulting in a strong impact of the official state ideology on the life of the schools and on the contents taught in many subjects,
- a total centralization of all decisions,
- a bureaucratic, non-professional management,
- a sequence of superficial reforms and disturbing interferences into the educational system motivated mostly by political or ideological reasons,
- the ideal of the "unified, uniform school" implying a severe suppression of almost all modes of differentiation (which itself was viewed and criticized as "elitism"),
- widespread mediocrity and loss of motivation among the more able pupils and teachers.

(b) All educational research institutions (ERIs) are trying to determine their new role within the changing educational structures (both as regards their activities and their funding), their staff often substantially diminished, and with new people being appointed into most of the leading positions in these institutions.

(c) The ERIs have started to establish working contacts with analogous institutions in Europe. Common fields of interest are being identified and possibilities of cooperation examined.

(d) The main constraints in the work of most ERIs are:
- lack of funds (the transition period from the centrally planned economy to a market economy is a very hard one for institutions like the ERIs),
- totally insufficient access even to basic books and journals in the field of educational research,
- insufficient foreign language skills of most researchers (the most widely spoken foreign languages being Russian, Hungarian, and German),
- insufficient technical equipment (lack of copy machines, faxes, access to Email, computers ...)

EDUCATIONAL RESEARCH AND THE PROBLEMS OF THE SECONDARY SCHOOL CURRICULUM

As pointed out in the San Marino paper the problems related to the curriculum currently play a significant role in the concerns of researchers. To understand this issue better, one has to bear in mind the main characteristics of our starting position (by starting position is meant the situation of November 1989. Since then much of what is here described has been profoundly changed):

- a highly centralized school system with a single curriculum obligatory for all schools of each particular type,
- very detailed curricula leaving almost no freedom to teachers as to what should be taught and how it should be taught. All the subject-matter was obligatory for all teachers and for all pupils,
- curricular documents containing too much ideology but, unfortunately, no clear and applicable statements of educational goals, no identification of the core curriculum, no description of minimal competencies, no setting of national educational standards. The documents were in most cases just lists of subject-matter items often embedded in an ideological background.

Presently a complicated multi-level process is under way: both the educational system and educational research are simultaneously undergoing thorough changes. In the process of planning and realizing various, often profound, changes in our educational system there is a strong need for relevant research results which could serve as a sound and clear base for all important decisions. Unfortunately, although there are not enough utilizable research results at hand, the decision makers cannot wait until educational research is able to present them with a clear picture because the public is not willing to wait so long to see visible changes in our schools.

It might well be the case that some of the problems currently being focused on in Slovakia might be of interest to other countries too and thus serve as a basis for cooperation. Therefore we give a list of them below. Before coming to that, however, let us add a terminological note to avoid possible confusion. The English term "curriculum" has quite a broad spectrum of meanings and it is sometimes a problem to find its equivalent in Slovak. In the present paper this term will be used either as referring to the content (the subject-matter) taught (in Slovak: *ufivo*) or to the document which defines and/or describes this content (in Slovak: *ufebne osnovy*). We do not use the term "curriculum" in its most general meaning of an entire educational programme comprising also the list of subjects and their time load (in Slovak: *ufebno plan*).

The curriculum as the content of the educational process

- What subject-matter should be taught within the particular subjects and at particular age-grade levels?

- Which theoretical principles or assumptions should underlie and guide the didactical transposition of the whole body of human knowledge into the school curriculum?

- Core curriculum - what is it? How (by what procedures) should it be determined? What is its role in the learning process? How does one find the optimal balance between regarding the core curriculum as a conceptual skeleton of the particular areas of human knowledge (this rather theoretical approach is typical e.g. of maths and natural sciences) and viewing it more pragmatically as a list of facts or skills which, from the point of view of society, every pupil should master in order to make his/her life and the life of society easier (or even possible)?

- Minimal competency - what is it (in general and in the particular subjects)? By what procedures should it be determined? What terms should be used to define it? What is its role in the education of the individual?

- National educational standards - what are they? How (by what procedure and according to which criteria) should they be determined? What terms should be used to define them? What is their role in education? How should their achievement by pupils and schools be measured?

- By what means should the curriculum reflect the particular educational needs of different groups of pupils? Should the main differences lie in the content (subject-matter)? Should it lie in the extent and intensity of learning (e.g. amount of lessons in a subject per week)? Or should it lie in the teaching methods?

- What is the optimal balance between the factual knowledge contained in the curricula and presented to pupils, and teaching pupils how to seek out relevant information when needed? "Should the pupils know which is the highest mountain in the world or should they merely know where to find the data when they happen to need it?"

- The traditional problem of the inter-subject relationships and the integrated curriculum. Our schools suffer from a traditional particularism which leads to a fragmentation of the curriculum into too many subjects, topics and sub-topics. The instruction is then not able to build and reveal the links between the contents taught separately in different subjects.

- Impact of modern technologies on the curriculum.

The curriculum as a document

Although much has already changed since 1989 in our educational system, the national curricula (ufebne osnovy) still remain the main official documents providing the teachers with information and guidance as to what should be taught in the schools. The curricula of most subjects do not specify only the concrete subject-matter, but also its distribution into grades, the sequence in which the particular topics should be presented, and the optimal amount of lessons which should be devoted to each of them. From this point of view the Slovak educational system is still therefore a highly centralized one. However, there is much discussion going on as to whether this should continue to be the case also in the future and, if not, what the new curricula should look like. The following are the focus questions of this debate:

- What structure should the curriculum of a particular subject have in its documental form?

- What kind of information should it provide for the teacher ?

- To what extent should curricula be obligatory? What is the optimal "degree of freedom" that should be given to schools and teachers?

- Which of the following should feature in the curricular documents:
    - aims and goals of the subject?
    - concrete subject-matter?
    - recommendations regarding the teaching methods and the organization of instruction?

- What is the optimal procedure for curriculum design? Who should be the authors? Who should be the reviewers? What should the roles of teachers, educational researchers and policy-makers in this process be? How can a permanent innovation of the curriculum be best organized?

- What other kinds of materials should accompany the curricular documents so that a possibly clear and sufficient information transfer from the curriculum designers to the teachers is guaranteed?

It is by no means our intention to claim that the problems listed above are new ones. Rather, we suppose that in most countries with a well-developed educational system these sorts of questions had to be raised and answered in some concrete way already many years ago. Obviously, all the questions were answered in some way in our country as well. We are however forced to reconsider many of those answers in the light of our new situation. We would greatly appreciate learning more about the different ways and approaches other countries have chosen, since learning from the experience of others would definitely make our own efforts more effective.

NOTE:

Czecho-Slovakia is a federation of two republics. This paper focuses on the situation in the Slovak Republic (though many of the general characteristics described apply to the whole country). The situation in the Czech Republic is described in more detail in the paper written by Walterova and included in this volume.

REFERENCE

Walterova, Eliska (1992): Research on Secondary School Curriculum in Czechoslovakia (Czech Republic). Document DECS/Rech (92) 19, prepared for the CDCC workshop in Valletta

NEW AIMS AND GOALS OF THE SLOVENIAN PRIMARY AND LOWER SECONDARY SCHOOLS

Darja Piciga
Educational Research Institute, Gerbiceva 62
61111 Ljubljana, Slovenia

The new general aims of primary and lower secondary school[1], based on the recent understanding of education, on the development of education in advanced countries, on the recommendations of International organizations and communities, and in tune with the expected trends in the educational system and the socio-economic characteristics of the Republic of Slovenia[2], could be defined as follows:

"The basic aim of primary education is to enable the development of the personality of an individual, which means a balanced development of different aspects of a child's personality and his/her potential, and to educate him/her for autonomous balancing of his/her needs with the needs of society. This aim is also accomplished by encouraging a child's self-confidence and by fostering a positive self-image.

Primary school provides a foundation for the further education and training of a young person, with emphasis on permanent education and quality knowledge for all young people. It develops basic classical and modern literacies, enables the understanding of natural and social phenomena, ensures the basic knowledge of languages and humanities, aesthetic experience, artistic expression and evaluation, and provides for the healthy physical and psychological development of a young person.

Primary school must help the child acquire and practise the democratic values of tolerance, cooperation, responsibility and respect for the rights of others. It prepares him for a life within a multi-cultural society with respect for its natural and cultural heritage. Primary school provides for the transmission of the national and universal heritage as well as for the development of national identity."

The following is a more detailed description of the aim:

- Offering equal educational opportunity is such an essential task of education that it is included in the constitutions of many countries. In the new Slovenian Constitution this right has not been specifically stated, therefore we must endeavour to make primary school legally responsible for ensuring the equality of educational opportunities, as well as providing for children and groups with special learning and psycho-social needs.

Equal educational opportunity is to be understood as a long-term aim to raise the general cultural, intellectual and educational level and, at the same time, reduce individual differences originating from socio-economic inequalities (but not reducing biologically conditioned differences).

- The basic aim of primary school is to enable the development of the personality of an individual and to educate him for an autonomous balancing of his needs with the needs of society. Thus primary school should foster a harmonious development of different aspects of the human personality; the physical, emotional, social, and moral, the development of cognitive abilities, skills, creativity, imagination, and other areas of personality development. It is important that primary school develops a child's self-confidence and fosters a positive self-image. It should also promote individual growth in tune with the values of the community, in consideration of and with responsibility towards others. In view of the present weaknesses of the Slovenian primary school and the overall social context, the development of those personality traits, attitudes, and abilities necessary to face and overcome problems should be emphasized.

- The basic aim of primary school education also implies quality education for all young people, education where talents and dispositions can fully develop and which is relevant to the social, cultural and economic needs of the Republic of Slovenia, as well as to the general needs of civilization.

- In view of the educational aims of primary school further education and training of a young person should be considered, emphasizing thereby the preparation for permanent (lifelong) education. Primary school should educate pupils to successfully face the developing economic and social needs of the Republic of Slovenia. It should enable the development of those personality traits and dispositions which lead to personal initiative, the spirit of enterprise and flexibility when meeting changes in their future employment and in other aspects of life.

The foundations for further education that are to be ensured by primary schools encompass the acquisition of knowledge, the development of skills, the formation of a positive attitude towards learning, and the acquisition of the techniques and methods as well as positive habits for individual and group learning and work. Creating suitable conditions for experiencing enjoyment in discovering and creating new knowledge is also an important task of a modernized school.

- One of the most important objectives is the development of a democratic society, which respects its cultural and natural heritage. Primary school should cultivate respect for one's own moral values and those of others, promote the transmission of the national heritage and develop national identity. This includes the following:

a. respect, cultivation and development of all racial, national, linguistic, cultural, ideological, religious, sexual, and personal differences which do not endanger peace and peaceful co-existence;

b. promotion of knowledge, skills, abilities, competencies, attitudes and values which will enable pupils to participate as active and informed citizens in the democratic Slovenian society within an international context. Children should be taught how to resolve conflicts peacefully. Primary school should foster peace and peaceful co-existence among nations and teach respect for human rights, basic liberties and principles;

c. cultivation of independent thinking enabling an individual actively to cooperate and responsibly contribute to the life of the community in the spirit of democratic principles;

d. presenting different answers to questions regarding the general philosophy of life, politics, social values, ethics and different life styles. In a way suitable to the pupils, primary school should prepare them for autonomous decision-making and taking a stand on the basis of critical judgment, which leads to the development of a free individual. This should be achieved through experiential learning of how to behave in social situations and through training for successful communication;

e. enabling the understanding of natural and social phenomena in all their complexity and inter-relationship, developing the responsibility of young people towards a balanced socio-economic progress which protects the environment and develops a treatment of nature which preserves it for future generations.

f. cultivation of the respect for and active understanding of values regarding the Slovenian identity, language, history and culture, and an awareness of the world's cultural heritage and the heritage of other nations and nationalities lying within the Republic of Slovenia.

- One of the essential objectives of primary education has always been the transmission of a basic understanding of one's immediate as well as wider natural and social environment and the development of basic literacies. Getting ready for the post-industrial society of the 21st Century requires a new attitude towards knowledge and literacy.

Primary school should therefore enable and promote the active acquisition of quality knowledge and develop the pupils' abilities and skills in different areas. Here knowledge is valued which is well-organized and includes not only facts, concepts and principles but also the mastery of various strategies needed for specific cognitive processes and in different areas.

Primary school should develop basic classical as well as modern literacies (listening, reading, writing, speaking and communication skills, media and visual literacy, computer and informational literacy, etc.)

Primary school ensures and enables the construction of:

- basic knowledge in mathematics,
- natural sciences, technical and technological knowledge and an understanding of the role of natural sciences and technology in a society.
- basic knowledge in social sciences, languages and humanities, emphasizing the knowledge of foreign languages (in the immediate future, the primary school should offer at least two foreign languages);

- The programme of primary school should not be limited to the acquisition of knowledge and mastery of skills, but attach equal importance to objectives such as:
  - the development of comprehension, understanding, experience, expression, creation and communication in different areas of
  - arts and culture; discovering pupils' dispositions and special talents in these areas as well as ensuring active participation in them;
  - encouraging physical and motor development as well as interest in manual skills, care for a healthy life and creative use of spare time.

In its upbringing role, primary school should cooperate with other education agents, especially parents.

NOTES

1. In Slovenian terminology, the term Primary Education covers both primary and lower secondary education (Ages 6 - 15).
2. New aims of the Slovenian primary school were constructed by the expert group:
   Darja Piciga, PhD, Educational Research Institute, Ljubljana (coordinator)
   Janez Becaj, MSc, Faculty of Philosophy, Ljubljana
   Dora Gobec, Educational Research Institute, Ljubljana
   Martin Kramar, PhD, Faculty of Pedagogy, Maribor
   Lidija Magajna, MSc, Consulting Centre, Ljubljana
   Barica Marentic-Pozarnik, PhD, Faculty of Philosophy, Ljubljana
   Cveta Razdevsek-Pucko, PhD, Faculty of Pedagogy, Ljubljana
   Marija Strojin, Consulting Centre, Ljubljana
   Veljko Troha, PhD Faculty of Pedagogy, Ljubljana
   Anica Uranjek, Consulting Centre, Ljubljana

POSTSCRIPT

Paul Heywood
Assistant Director of Education
Education Department, Floriana, Malta

It may not be amiss to reflect briefly on the immediate impact of the workshop on the thinking of an administrator caught up in the day-to-day management of the Upper Secondary sector at a time of rapid expansion.

There is a strong commitment in Malta to promoting interdisciplinarity at Sixth Forms and the University. Some interesting experiments are in progress, e.g. the Systems of Knowledge project and the University Foundation Course. The smallness of our system facilitates bold innovatory approaches on a national scale. Indeed efforts are being made to contain the requirement of interdisciplinarity within bounds that help define this important yet elusive quality, thus rendering it more manageable and therefore more meaningful to both teachers and students.

The Council of Europe workshop, despite the misgivings voiced by a few participants, has confirmed us in our resolve to consolidate the present schemes for promoting interdisciplinarity, which eventually might prove of general interest.

Liv Mjelde's paper on the relationship between the world of work and that of the school, between intellectual and manual labour touched a vital issue and a very sore point in the Maltese educational set-up. It also sheds light on the deep social divide between the traditional professions and those fathered by modern engineering technology, something which Malta has inherited from the British system.

We are chipping away at this barrier. For a start, students between the ages of 16 and 19 seeking admission to University are obliged to work on a Technology Project, involving the 'application of knowledge for making and doing purposeful and useful things'. Incidentally, technology starting from scientific knowledge, combines manual and mechanical dexterity, creative design, testing, note-taking, decision-taking, calculated risk, evaluation, etc., implying an interdisciplinary approach to the solution of a practical problem.

For the "twain to meet" it is felt that the potential intellectual elite, the university student, whatever her/his bent and interests, needs to gain some insight into modern technology.

The assessment of these technology projects is school-based and students are free to seek help from any quarter. This explains why they are called upon to give an oral exposition to the moderator when it is seen they are not giving sufficient proof of understanding the technological implications of their efforts. Since even parents get involved in the exercise we should like to hope that a quiet social revolution is being sown.

Another theme of central interest to the Maltese delegation was broached in one of the workshops and taken up in informal discussions. What is taught in school is not absolutely relevant to later life and is imparted in too abstract a way; it is not based on the student's personal experience. This is aggravated by the impersonal teaching style adopted by many teachers, especially in Maltese secondary schools. Curriculum overload, of course, adds pressure on the teacher to deliver her/his subject in a formal and sterile lecture-like fashion.

These methods contrast strikingly with those adopted in adult education. There the emphasis on personalised learning and on dialogue between learner and facilitator, with the former in a position jointly to steer courses, helps mature the adult educator. Now since most Maltese facilitators happen to be also mainstream teachers, they are learning to step down off their classroom platforms and to shed their predominantly magisterial mode in order to adopt more participative methods of teaching, affording pupils a better chance of developing important skills like self-expression and teamwork. This healthy spin-off effect of adult education on secondary schooling is already being felt in our system. It is perhaps another theme that might be taken up in future workshops.

These features, among others in the Maltese educational set-up, are slowly modifying the long-standing framework structured along traditional academic disciplines.

As for research, it is mainly action research, the number of projects being far from negligible. Indeed it is unfortunate that we have not yet found the wherewithal for coordinating these various individual and small-group initiatives. Even then, however, urgent changes and adjustments dictated by ever-accelerating social and economic change in a small open economy like Malta's will often outpace both the researcher and the planner, while the administrator has to act in spite of everything to meet immediate basic needs.

LIST OF PARTICIPANTS/
LISTE DES PARTICIPANTS

**CHAIRMAN, RAPPORTEUR
AND LECTURERS/ PRESIDENT,
RAPPORTEUR ET
CONFERENCIERS**

Dr. Paul Heywood
(Chairman/President)
Ministry of Education and Human
Resources, Floriana.
Malta

Dr. Clare Burstall
(Rapporteur Général)
Director NFER
The Mere, Upton Park
GB-Slough, Berkshire SL1 2DQ

Prof. Jacques Colomb
Directeur du département
"Didactique des Disciplines"
INRP, 29 rue d'Ulm
F-75230 Paris-Cedex 5

Mr. Gábor Hálasz
National Institute for Public
Education
Dorottya u.8, H-1051 Budapest

Prof. Jean Ruddock
Director, QQSE Research Group,
Division of Education,
University of Sheffield, PO Box 597,
Hicks Building,
Hounsfield Road
GB-Sheffield S10 2UN

Dr. Johan Van Bruggen
Secretary, CIDREE, Boulevard 19453
7511 JW Enschede PO Box 2041
NL-7500 CA Enschede

Prof. Kenneth Wain
Head, Department of Educational
Studies
Faculty of Education
University of Malta
M-Msida

Prof. Liv Mjelde
Statens Yrkespedagogiske Hogskole,
Postboks 2803 Toyen
N-Oslo 6

DELEGATES/DELEGUES

Albania/Albanie
Ms. Irena Vangjeli
Inspector of Secondary Education
Ministry of Education
Tirana

Austria/Autriche
Dr. Herbert Puchta
Pädagogische Akademie des Bundes
in der Steiermark
Hasnerplatz 12
A-8010 Graz

Belgium/Belgique
Mr Jean Ravez
Inspecteur Général de l'enseignement
Secondaire
Commissariat Général aux Relations
Internationales de la Communauté
Francaise de Belgique
65, avenue Louise - Boite 9
B-1050 Bruxelles
Privé: avenue d'Hyon 142
B-7000 Mons

Cyprus/Chypre
Mr Christos Georgiades
Inspector of Secondary Education
Ministry of Education
Cy-Nicosia

Czechoslovakia/Tchecoslovaquie
Dr Eliska Walterova
Institute of Educational and
Psychological Research
Faculty of Education, Charles
University
Rettigové 4
CS-116 39 Prague

M. Vladimir Burjan
Pedagogical Research Institute
Kutlikova 17
CS-825 55 Bratislava

France
Prof. Jacques Colomb
Directeur, Département "Didactiques
des Disciplines"
INRP
29 rue d'Ulm
F-75230 Paris Cedex 05

Germany/Allemagne
Dr Happ
Leiter des Staatsinstituts für
Schulpädogogik und
Bildungsforschung
Arabellastr. 81
D-8000 München

Ireland/Irlande
Ms. Sandra Ryan
Educational Research Centre
St Patrick College
IRL-Dublin 9

Italy/Italie
Mme Michela Mayer
CEDE
Centro Europeo dell'Educazione
Université de Rome
Villa Falconieri
I-00044 Frascati (Rome)

Malta/Malte
Dr. Paul Heywood
Ministry of Education and
Human Resources
Floriana

Prof Kenneth Wain
Head, Department of Educational
Studies
Faculty of Education
University of Malta
Msida

Dr James Calleja
Executive Director
Foundation for International Studies
University of Malta
St Paul Street
Valletta

Mr Frank Ventura
University of Malta
Msida

Mr Joseph Fenech
University of Malta
Msida

Dr Mary Darmanin
University of Malta
Msida

Mr Michael Sant
University of Malta
Msida

Ms Josephine Cilia
Department of Education
Floriana

Mr Joe Agius
Department of Education
Floriana

Mr Andrew Buhagiar
Department of Education
Floriana

Mr Joe Sammut
Department of Education
Floriana

Mr Edgar White
Department of Education
Floriana

Mr Frank Gatt
Department of Education
Floriana

Mr Arthur Sammut
Department of Education
Floriana

Mr Joe Zammit Mangion
54 St Mary Street
Sliema

Mr Charles Mizzi
Department of Education
Floriana

Norway/Norvege
Prof Torstein Harbo
University of Oslo
Institute for Educational Research
P.O. 1092 Blindern
N-0317 Oslo

Poland/Pologne
Ms Bozena Chrzastowska
Instytut Filologii Polskiej
ul. Al. Niepodlegrosci 4
PL-61-874 Poznan

Romania/Roumanie
Mr. César Birzea
Director of the Institute of Educational
Science
37 Stirbei Voda Street
RO-70732 Bucharest

Slovenia/Slovenie
Dr Darja Piciga
Project director for primary, secondary
and science education and
cognitive development
Educational Research Institute
Gerbiceva 62
SLO-61000 Ljubljana

Spain/Espagne
Mr Vicente Riviere
Gabinete de la Secretaria de Estado de
Educacion
Ministerio de Educacion y Ciencia
Alcala, 34
E-28014 Madrid

Switzerland/Suisse
M. Jean Pierre Meylan
Secrétariat Général de la CDIP
Sulgenechkstrasse 70
CH-3005 Bern

United Kingdom/Royaume-Uni
Prof. Jean Ruddock
Director, QQSE Research Group
Division of Education
University of Sheffield
PO Box 597, Hicks Building
Hounsfield Road
GB-Sheffield S10 2UN

OBSERVERS/OBSERVATEURS

Consortium for Development and Research in Education in Europe (CIDREE)
Dr Johan C. Van Bruggen
Secretary
CIDREE
Boulevard 1945 3
PO Box 2041
NL-7511 JW Enschede

WCOTP/CMOPE
World Confederation of Organizations of the Teaching Profession
Confédération Mondiale des Organisations de la Profession Enseignante

FIPESO
International Federation of Secondary Teachers
Federation Internationale des Professeurs de l'Enseignement Secondaire Officiel

Mr Alfred Buhagiar
Malta Union of Teachers
Teachers' Institute
213 Republic Street
Valletta

UNESCO
excused/excusé

SECRETARIAT

Council of Europe
Dr Michael Vorbeck
Chef de la Section de la Documentation et de la Recherche Pédagogique
Conseil de l'Europe
Strasbourg CEDEX
Poste 24.00
Tel:(33) 88 61 49 61
Fax: (33) 88 36 70 57

Madame Daniele Imbeck
Secretariat de M. Vorbeck

Malta Secretariat
Dr Paul Heywood
Ass. Director of Education
Education Department
Floriana
Malta
Tel: (356) 239842
Fax: (356) 221634

Coordination
Dr James Calleja
Executive Director
Foundation for International Studies
St Paul Street
Valletta VLT 07
Tel: (356) 234121/2
Fax: (356) 230551

Assistants
Ms Margaret Abdilla
Assistant/Secretary

Ms May Pantalleresco
Conference Executive

Ms Rose Anne Zammit
Research Assistant

Mr François Grech
Typesetting

Milton Keynes UK
Ingram Content Group UK Ltd.
UKHW031133141024
449569UK00006B/230